Expotecnologia 2022: Research, Innovation and Development in Engineering (Expotecnologia 2022)

IOP Conference Series: Materials Science and Engineering
Volume 1299

Cartagena, Colombia
2 - 5 November 2022

Editor:

Libis Valdez Cervantes

ISBN: 978-1-7138-9255-7
ISSN: 1757-8981

Printed from e-media with permission by:

Curran Associates, Inc.
57 Morehouse Lane
Red Hook, NY 12571

Some format issues inherent in the e-media version may also appear in this print version.

Institute of Physics
Dirac House, Temple Back
Bristol BS1 6BE UK

Phone: 44 1 17 929 7481
Fax: 44 1 17 920 0979

techtracking@iop.org

Additional copies of this publication are available from:

Curran Associates, Inc.
57 Morehouse Lane
Red Hook, NY 12571 USA
Phone: 845-758-0400
Fax: 845-758-2633
Email: curran@proceedings.com
Web: www.proceedings.com

TABLE OF CONTENTS

IOP Conf. Series: Materials Science and Engineering 1299 (2024) 011001 doi:10.1088/1757-899X/1299/1/011001

Expotecnología 2022

Organized by

Sponsors:

IOP Conf. Series: Materials Science and Engineering 1299 (2024) 011001 doi:10.1088/1757-899X/1299/1/011001

Expotecnología 2022 was an event that took place in the city of Cartagena de Indias, from November 2 to 6, 2022, with the slogan " The future of engineering in the face of technological disruption " This eighteenth version was a meeting that brought together the scientific and academic community in spaces for the dissemination and transfer of knowledge, where innovation was the essence and a fundamental part of the event. Expotecnología was created in the Faculty of Engineering Sciences - FACI, of the Antonio de Arévalo University Foundation - UNITECNAR and in this version it maintained as mission, the development of the interaction between science, technology and society, through establishing of knowledge networks and academic and professional cooperation. The main objective was to address the challenges that emerge from the everyday life of the engineering community, the problems immersed in the development of projects and programs of Higher Education Institutions (IES), organizations and research centers at the national and international level. In the development of the day of presentations in a virtual way, it was possible to promote the dissemination of the research work of professionals, teachers and students from all areas of engineering, framed in four (4) thematic axes, as follows:

- New technologies: Big data, Artificial Intelligence, Business Intelligence, Software Engineering, Education and new technologies and Mobile Applications.
- Petrochemicals and process control: Materials Science, Instrumentation and Control Systems, Robotics and Automation and Industrial Maintenance.
- Electricity, electronics and telecommunications: IoT, Fiber optics, Digital Signal Processing, Wireless Networks, Mobile Communications, Power Systems, Energy Storage, Smart Grids, Renewable Energies, Operation and Distribution, Energy Dispatch, Energy Markets
- Construction Summit: Modern Building Systems and Computational Modeling.

On this occasion, the event was attended by 512 people, representatives from 25 national and 6 international universities. 32 presentations were made, of which 32 articles were selected for this volume of the IOP Conference Cycle: Materials Science and Engineering and 13 articles for the book Engineering, Innovation and the fourth industrial revolution., After a process of Peer Review. The plenary conferences were in charge of 2 national and 8 international speakers .

Invited speakers:

PhD. ING. JAIME ESTEVA
PhD. GABRIEL ROMÁN MELÉNDEZ
PhD. JESÚS DAVID ANGULO REYES
PhD. ROBERTO GIORDANO LERENA
PhD. AARON RODRIGUEZ
PhD. SAMUEL CUADRADO BELLIDO
PhD. ARYANNE QUINTAL
PhD. VIANNEY LARA PRIETO
PhD. MARÍA ILEANA RUIZ CANTISANI
PhD. SONIA HELENA CONTRERAS ORTIZ
MSC DANIEL RENGIFO ALMANZA

Organizing Committee

President

Msc. Libis Valdez Cervantes, Antonio de Arévalo University Foundation

Editorial Comité

Msc. Yesid Tarriba Lezama, Antonio de Arévalo University Foundation

IOP Conf. Series: Materials Science and Engineering 1299 (2024) 011001 doi:10.1088/1757-899X/1299/1/011001

Technical Comité

Msc. Luz K Ahumada, Antonio de Arévalo University Foundation,
Msc. Yesid Tarriba Lezama, Antonio de Arévalo University Foundation
Msc. Mayerlis Vasquez, Antonio de Arévalo University Foundation
Msc. Libis Valdez Cervantes, Antonio de Arévalo University Foundation

Scientific Committee

Coordinator

Msc. Yesid Tarriba Lezama, Antonio de Arévalo University Foundation

Special thanks

We greatly thank all the authors for their valuable participation in Expotecnología 2022 and all the peer reviewers for their commitment and dedication in reviewing the papers to be published in IOP and Book chapters.

1. Number of applications received: 36
2. Number of applications reviewed: 34
3. Number of Accepted Applications: 32
4. 14 National institutions that supported (UTS, Unisinu, ITSA, technological Comfenalco, Uniguajira, American University, Free University, Creainn, University of Pamplona, ECCI University, Uniminuto
5. 2 international institutions (New York Institute of Technology, Autonomous University of Hidalgo)
6. Connections Via Zoom Renata 120.
7. 10 International speakers

Msc. Yesid Tarriba Lezama
Editors of the Proceedings of EXPOTECNOLOGIA 2022

IOP Conf. Series: Materials Science and Engineering 1299 (2024) 011002 doi:10.1088/1757-899X/1299/1/011002

Peer Review Statement

All papers published in this volume have been reviewed through processes administered by the Editors. Reviews were conducted by expert referees to the professional and scientific standards expected of a proceedings journal published by IOP Publishing.

- **Type of peer review:** Double Anonymous
- **Conference submission management system:** Morressier,IOP
- **Number of submissions received:** 13
- **Number of submissions sent for review:** 13
- **Number of submissions accepted:** 13
- **Acceptance Rate (Submissions Accepted / Submissions Received × 100):** 100
- **Average number of reviews per paper:** 2
- **Total number of reviewers involved:** 16
- **Contact person for queries:**
  Name: Yesid Tarriba Lezama
  Email: expotecnologia@unitecnar.edu.co
  Affiliation: Unitecnar

IOP Conf. Series: Materials Science and Engineering 1299 (2024) 012001 doi:10.1088/1757-899X/1299/1/012001

Carbon footprint measurement using a mobile app integrated with IoT technology

L. Blanquicett[1], L. Murillo[2], F. Fernandez[3]
1 Professor, Universidad del Sinú - Cartagena, Colombia
2 Professor, Universidad del Sinú - Cartagena, Colombia
3 Professor, Corporacion Universitaria Rafael Nuñez - Cartagena, Colombia

E-mail: lblanquicett@unisinucartagena.edu.co

Abstract. It is known that global warming and the carbon footprint are directly proportional, that is, the more carbon footprint there is, the more global warming there will be, for this reason it was decided to take this environmental problem and calculate the carbon footprint of the engineering faculty of the Sinú University, in this way, the university can know the CO2 emission that is emitted in the faculty and in this way be able to take actions regarding the mitigation of these emissions. For this, a mobile application was developed with the Dart development language, this application can perform manual calculations of the carbon footprint, and a real-time calculation was implemented through a sensor that is connected to an Arduino, this Sensor can measure the electrical flow of a device and sends the information to an IoT server. Knowing the electric flux, the carbon footprint can be calculated. The application allows to control emissions, and organizes this information by means of graphs, and in this way the user can have a better understanding of the stored information.

1. Introduction

Currently society is going through an era of environmental crisis, temperatures are getting higher, leading to the melting of the geographic poles. The impact has been increasing over the last few years, bringing great consequences and side effects. This is due to the phenomenon called "greenhouse effect", which is created when some gases retain the energy emitted by the earth after being heated by solar radiation.

Greenhouse gases (GHGs) are increasing significantly according to the report presented by IDEAM and UNDP, prepared in the country, the increase has been 15%, going from 245 million tons of CO2 in 1990 to 281 million tons in 2010, causing solar radiation to remain within the atmosphere for longer than necessary. In this way, the greenhouse effect is not in itself a problem, but the excessive actions of mankind are making it difficult to eradicate, it is causing it to be further accentuated by the increase in gas emissions, especially carbon dioxide (CO2), which raises the planet's temperature.

It is worth highlighting the impact of the industrial revolution on the climate, which has led to realize hundreds of investigations that indisputably show that global warming is closely related to this issue. For example, the Free University of Berlin, which is one of the pioneering institutions in these investigations, reported that it is the main cause, based on the pioneering study of the "natural climate archive" of the last 500 years.

IOP Conf. Series: Materials Science and Engineering 1299 (2024) 012001 doi:10.1088/1757-899X/1299/1/012001

2. Literature review

With the accelerated and diversified development of technologies in order to facilitate the most difficult tasks of the human being, these have begun to be introduced in the field of environmental care and even in the specialized area of carbon footprint calculation, as climate change is one of the biggest concerns and one of the most important problems currently faced by humans because many of the products consumed or used, including day-to-day activities largely denote the high index of greenhouse gases. Having said that, some investigations that propose solutions related to CO2 emission and carbon footprint were reviewed, including the following.

Internet of Things system for monitoring protected crops, it is an Internet of Things (IoT)-based project to monitor crops in specific areas, using sensors which will send information to a web server, these sensors collect different data, such as relative humidity, temperature, water volumetric content in the soil and other factors. The architecture of the project is a sensor-cloud-server application model, the data is automatically updated, and the user will be able to display it through the website or from the mobile device either through graphs or tables. This system can send alerts to users so that in this way they have full control of the crops in the greenhouses (Gómez et al., 2018).

SGreenH-IoT: IoT Platform for Precision Agriculture, it is a platform designed to meet the needs of remote monitoring, thus reducing the investment of human resources, the platform is focused on the field of precision agriculture, it has mechanisms that generate a balance between agricultural production and optimization of the resources used like water and fertilizers. This type of agriculture contributes to combat epidemic diseases, to optimize resources and also provides added value to agricultural production, the platform has a four-layer architecture, it has sensors which are responsible for collecting information regarding fields of crops and greenhouses, the information is sent by the ZigBee protocol to the central server in the cloud, this stored data is analysed in the cloud, showing statistics about the crops and external factors, and the last layer allows users to view all the information on the dynamic website and the farmer can activate the systems manually or automatically (Guerrero et al., 2017).

Green Multimedia: Informing People of Their Carbon Footprint through Two Simple Sensors, this project carried out in Ireland by Dublin City University students, Aiden R. Doherty, Zhengwei Qiu, Colum Foley, Hyowon Lee, Cathal Gurrin, Alan F. Smeaton aims to estimate CO2 emissions related with the transport of a person using a simple portable accelerometer, present on many mobile phones, used as a means of estimating the CO2 of an individual related to transportation.

It was achieved by building an algorithm to detect when an individual is driving, based on the values of the motion sensor x, y, z, and from those calculations, the time that is driven and stops, then it is possible to make an estimation of how many litres of gasoline are consumed and what can finally be converted into CO2. In addition, they conducted extensive research to provide an estimate of transport-related CO2 emissions through an interactive website and mobile application, which encompasses a set of users to be aware of their CO2 emissions (Doherty et al.., 2010).

3. Materials and methods

Throughout the work, different sensors were used, each one with the functionality we needed to perform the calculation of the carbon footprint in real time and thus be able to display the results from the mobile application. The sensors used during the work were as follows: Arduino UNO, ESP8266, RTC DS1307, ACS712, as shown in figure 1.

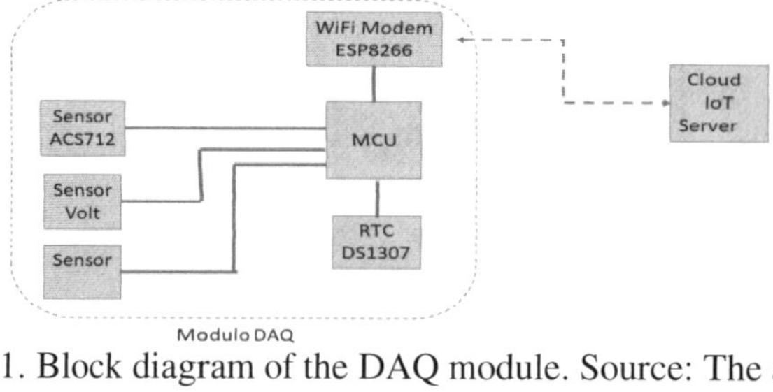

Figure 1. Block diagram of the DAQ module. Source: The authors

After making the connections, as shown in figure 2, the system was programmed so that every certain time they would send the data to the IoT server and thus be able to process the data that would be displayed in the application later.

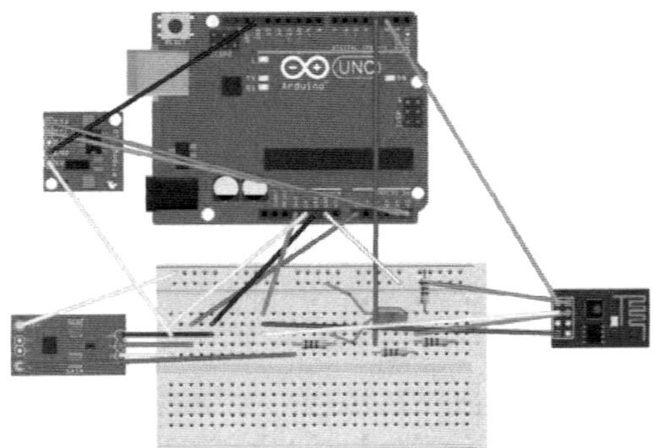

Figure 2. Graph of the result. Source: The authors

The carbon footprint is measured in CO2 emissions and in order to find the emissions, mathematical operations are necessary, each emission source has an emission factor, from the beginning of the work we knew that two types of calculations would be implemented, one manual and the other which would be carried out in real time through sensors capable of measuring the electrical flow and in this way being able to calculate the carbon footprint of the device that has been connected to it.

For the calculations, the parameters for calculating the GHG inventory were used as shown in table 1.

Table 1. Equivalent emission factors by activity in kg of CO2.

Activity			Emission factors of CO2
Diesel Consumption			10.2765 kg co2 emission/gal
Natural Gas Consumption			1.9801 kg co2 emission/m3
Electricity Consumption			0.221 kg co2 emission/KWh
Air transport	Domestic flights		0.205 kg co2 emission/km
	International flights	Short	0.116 kg co2 emission/km
		Long	0.135 kg co2 emission/km
Water Consumption			0.1427 kg co2 emission/m3
Paper Consumption			1.30 kg co2 emission/kg
Recycled waste generation	PET		2.538 kg co2 emission/kg waste
	Iron		2.9 kg co2 emission/kg waste
	Polypropylene		1.343 kg co2 emission/kg waste
	Glass		0.04 kg co2 emission/kg waste
	Aluminum		9.13 kg co2 emission/kg waste
	Paper + Cardboard		0.55 kg co2 emission/kg waste
	Plastic		0.035 kg co2 emission/kg waste
	Cardboard cups		10.15 kg co2 emission/kg waste
Dangerous waste generation	WEEE**** (Waste Electrical Electronic Equipment)		0.0846 kg co2 emission/kg waste
	biosanitary and uncategorized waste		0.0502 kg co2 emission/kg waste
	solvents		0.0991 kg co2 emission/kg waste
	contaminated containers and materials		0.0262 kg co2 emission/kg waste
	Acid, bases and solutions		0.0677 kg co2 emission/kg waste
	oils		0.0347 kg co2 emission/kg waste
	pesticides		7.37 kg co2 emission/kg waste
	organic waste		0.365 kg co2 emission/kg waste
	toner		2.76 kg co2 emission/kg waste
Industrial sewage			5.63 kg co2 emission/DQO
Built infrastructure			520 kg co2 emission/m2 built

To demonstrate that the manual calculation of the carbon footprint in the application works correctly, the manual calculation is carried out through a spreadsheet in Microsoft Excel software, the emission sources that were calculated were paper consumption and consumption of cardboard cups, this checking is performed taking the month of March as a reference. The results are shown in figure 3.

The measurement was carried out weekly starting on Monday, 02/March/2020 and ending on Tuesday, 31/March/2020, performing the calculations every Friday of each week and in the last week it was carried out on Tuesday, in the figure you can see the total carbon footprint that was calculated weekly. This same simulation was conducted from the mobile application and the results are totally matched with those of Excel, if the manual calculation of the carbon footprint from the mobile application works correctly.

The calculations made from the application must be stored in the database it is necessary for the user to save the results to be able to keep a detailed control of them, these calculations are displayed in a graph, which allows the user to better interpret these data. The results are shown in figure 3.

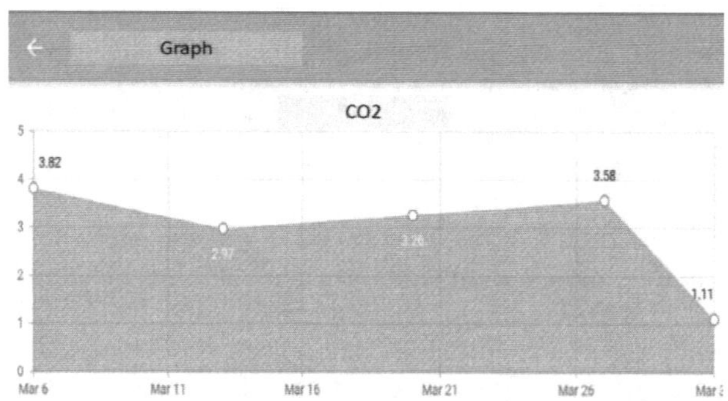

Figure 3. Graph of the result. Source: The authors

Calculation of the carbon footprint in real time. This type of calculation was carried out with the sensors previously described, each sensor with its respective functionality allowed to measure the electrical flow of the device to which it has been connected, for the work, a LED diode was used, as can be seen in Fig 4, in this way, the ACS712 sensor capable of measuring the electrical flow, sends the data to the IoT server via the ESP8266 Wi-Fi microcontroller module, this data is sent promptly at 23:59:55, thanks to the RTC DS1307 module to which the time was configured (GMT-5) and this data can be displayed from the application in the form of graphs so that the user can better interpret this information.

The deployment diagram illustrates the physical relationships of the different nodes that make up the calculation of the carbon footprint and the distribution of the components over said nodes. The TCP / IP and HTTP / HTTPS protocols are used. The site is hosted on a server with a Firestore database, as shown in figure 4.

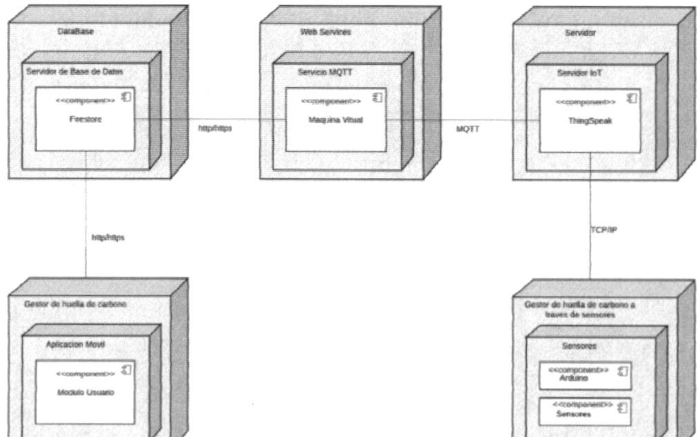

Figure 4. Deployment diagram. Source: The authors

The application provides these results graphically, the user being able to better interpret these data, the application also provides additional information when selecting the line of the graph, this will provide the user with the result of the carbon footprint and the date with which this calculation was stored in the database. The data is represented in the application as follows on figure 5.

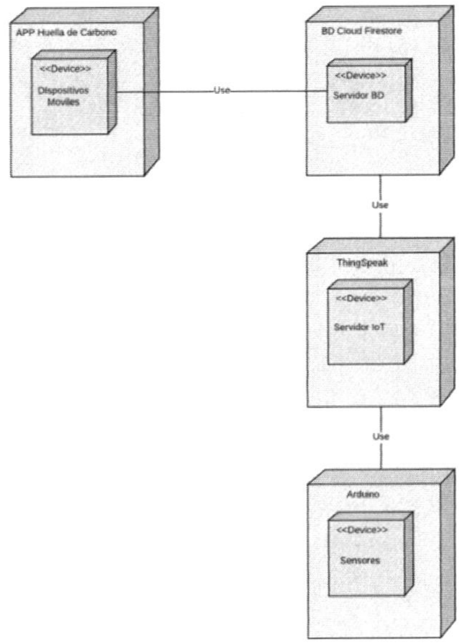

Figure 5. System distribution diagram. Source: The authors

4. Results and discussions

This work allowed the elaboration and construction of a prototype of a mobile application, which contributes to the real-time measurement of the carbon footprint that occurs in each unit and equipment of a building, in this case of the engineering faculty of the Universidad del Sinú mainly, in addition, the ease of knowing the carbon footprint that is being emitted in it. For the verification, data were taken manually from periodic measurements of the energy consumption of the installation site which were compared with the measurements made by the DAQ module system and the application.

These measurements were compared obtaining errors of less than 2%, which are partially attributable to manual and automatic mediations not being synchronized exactly over time. Additionally, there is a difference due to the resolution of each meter and some other part of the error is attributable to the calibration and measurement uncertainty. However, in the medium-term tracking (one week), it is noted more detail in the automatic measurement due to how often the data is sample. This allows for a more reliable measurement of energy consumption and its carbon footprint than manual measurement because this is estimated in longer periods of time, which is not suitable in environments where consumption undergoes variations during use. The application according to the tests of use with various users shows ease of use, the results are easily accessible, and their display and interpretation is adequate.

This system allows the integration of multiple sensors for reading electrical consumption in various units or by equipment in each unit, other sensors of the same type can be integrated practically without changes, only a few configurations are required in the DAQ module and in the application. Other sensors that allow the measurement of other variables can also be integrated into the calculation of the carbon footprint, such as: drinking water consumption, garbage, and waste production, etc. Additionally, it is possible to monitor the temperature and humidity in the units to determine the relationship between the consumptions of the air conditioning equipment and the thermal comfort profiles, to evaluate their efficiency during the operation of this. All these systems require the activation of calculation modules in the application and the addition of hardware modules to the DAQ and the

IOP Conf. Series: Materials Science and Engineering 1299 (2024) 012001 doi:10.1088/1757-899X/1299/1/012001

activation of software routines in the measurement, which were not enabled during the tests of this prototype, but which can be activated for more extensive tests.

Figure 6. Application screens. Source: The authors

The measurement of the environmental temperature and relative humidity variables is included, as an additional value, which allows future relationships between energy consumption and environmental conditions to be obtained, as part of the analysis of the carbon footprint of these activities. For this, we use a DHT11 sensor connected to the data acquisition module (DAQ) developed from an Arduino Uno and an ESP8266, in communication with the ThingSpeak server and shown in the mobile application developed in Kodular, the update times of the Data was initially set in 15 seconds, but can be modified in the app. The results are shown in figure 7.

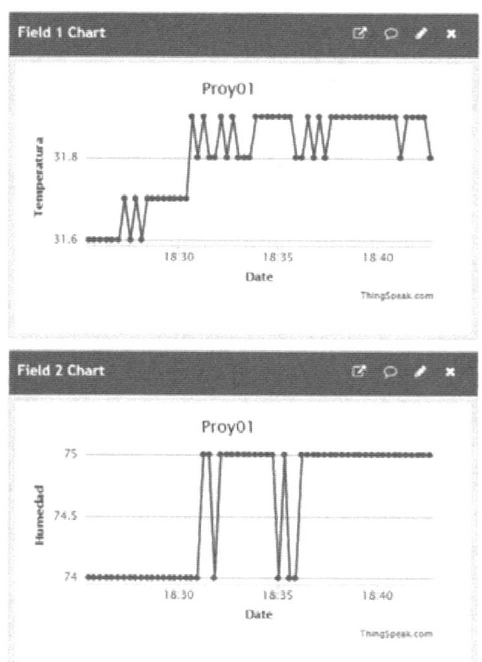

Figure 7. Manual calculation in Excel. Source: The authors

5. Conclusion

The application achieved all the objectives that were outlined from the beginning of the work, thus providing the Universidad del Sinú the possibility of having full knowledge about the carbon footprint and CO2 emissions, and thus, being able to take steps to mitigate the carbon footprint, and in this way the university can keep track of the devices that emit the most CO2 and find a way to prevent the excessive use of these devices.

The application will provide the university with the knowledge of the carbon footprint that is emitted in real time and manual calculations can be performed, the application also provides to the user with control of all the emissions that are stored in the application and thus be able to apply steps to mitigate the carbon footprint and corroborate with the help of the application, whether the steps are successful or not.

Acknowledgments

The authors express his sincere thanks to the Universidad del Sinú, Colombia for providing financial support for this research.

References

[1] A. Doherty, Z. Qiu et al (2010). Green multimedia: Informing people of their carbon footprint through two simple sensors. MM '10: Proceedings of the 18th ACM international conference on Multimedia, Vol. 1, Fiorenze, pp. 441-450.
[2] J. Gómez, S. Castaño, T. Mercado, A. Fernandez, J. Garcia (2018). Sistema de internet de las cosas (IoT) para el monitoreo de cultivos protegidos. Ingeniería e Innovación, 5(1).
[3] J. Guerrero, F. Estrada, M. Medina (2017). SGreenH-IoT: Plataforma IoT para Agricultura de Precisión. Sistemas, Cibernética e Informática, Vol. 14, No. 2, pp. 53-58.
[4] X. Li, H. Tan, A. Rackes (2015). Carbon footprint analysis of student behavior for a sustainable university campus in China. Journal of Cleaner Production, 106, 97-108.
[5] D. Manso, C. Parrado, A. Aristizabal (2017). Inventario de gases efecto invernadero en la Universidad de Bogotá Jorge Tadeo Lozano (Utadeo). Revista Mutis, 7(2), pp. 44-58.
[6] O. Robinson, A. Tewkesbury, S. Kemp, I. Williams (2017). Towards a universal carbon footprint standard: A case study of carbon management at universities. Journal of Cleaner Production, in press, doi: http://dx.doi.org/10.1016/j.jclepro.2017.02.147
[7] Secretaría Distrital de Ambiente. Alcaldía Mayor de Bogotá (2015). Guía para el cálculo y reporte de Huella de Carbono Corporativa, Bogotá, D.C., pp. 9-37.

A Deep Learning Approach to Classification Pneumonia in Thorax Images

Eugenia Arrieta Rodríguez[1], Agustin Naar[2] and Margarita Gamarra[3]

[1,2]Department of Systems Engineering, Faculty of Engineering, University of Sinú Cartagena, Av el bosque Cartagena,Colombia; investigacionsistemas@unisinucartagena.edu.co
[3]Departament of Computational Science and Electronic, Universidad de la Costa, CUC, Barranquilla 080002, Colombia; mgamarra3@cuc.edu.co

*Corresponding author: investigacionsistemas@unisinucartagena.edu.co

Abstract. An algorithm of automatic learning was developed. That is abke to identify radiographic images with a pneumonia and not pneumonia diagnose based on a data set published in the "Kaggle" platform by the Radiological Society of North America, from which we obtained a set of specific images with their labels that were divided by 70% to train a convolutional neuronal network model consisting of two convolutional layers for the extraction of characteristics in each image, and ends with a classification stage in the training of the model, to conclude with 74% in metrics of accuracy given by test tests with 30% of the data set.

1. Introduction

Pneumonia is an acute infection of the lung parenchyma which can cause from inflammation to accumulation of fluid in the lungs, its etiology may be due to infection by viruses, bacteria or fungi.

It affects non-hospitalized and hospitalized patients, it is characterized by a series of clinical manifestations such as the appearance of fever and/or respiratory symptoms (cough, expectoration) among others, together with the presence of changes in the chest X-ray, it represents more than 15% of all deaths of children under 5 years of age globally[1].

In Colombia, according to Vital Statistics (DANE), in 2016 there were 7,570 deaths, whose cause of death (ICD-10) was Pneumonia, unspecified organism and 24 due to Bacterial Pneumonia, not elsewhere classified [2].
According to the Hospital Clínico Universitario de Santiago de Compostela (Spain) in the journal Neumo Experts in Prevention, it is estimated that each year, 150 million children develop the disease and 11 million children are hospitalized due to pneumonia, and almost all of them live in developing countries, such as Colombia [3][4]. As part of solutions to this type of problems on a global scale, technology has made a significant contribution to combat diseases where Artificial Intelligence (AI) is the protagonist because it proposes the combination of algorithms with the purpose of creating

machines that be of help and present the same capacities of the human being. Now, it is well known in the Health sector that Radiological Imaging (CXR) is used for the diagnosis of Pneumonia. However, currently projects focused on the study and/or prediction and data processing systems can help in the diagnosis. of Pneumonia, thus, advanced technology proposes the use of Neural Networks that belong to the discipline of Artificial Intelligence (Artificial Neural Networks (RNA), inspired by biology. The name given to the type of Neural Networks that are Processing images are Convolutional Neural Networks (Convolutional Neural Network "CNN") which are necessary to generate a prediction model to help provide solutions to a problem in charge in this case of the health area, specifically pneumonia.

Convolutional Neural Networks is implemented in this research project to obtain significant features on a set of CXR images of patients with and without pneumonia, which in turn were used to train a mathematical model with the ability to generalize predictions from the different pneumonia images. This type of training in the field of artificial intelligence is recognized as supervised learning, where the training algorithm is provided with a set of data and its corresponding labels (Labels) that denote what certain particular data corresponds to. It is about obtaining greater precision in the diagnosis and in the results so as to allow timely intervention and invasion in the body, improve the quality of care, provide greater comfort to patients and provide greater access to technology services in Medicine.

It is clear that Artificial Intelligence is not going to replace doctors, but it is of great help when making decisions in the diagnosis of pneumonia. It also helps to reinforce the knowledge of general practitioners who do not have enough experience as radiology specialists when interpreting a chest X-ray.

Although there are enormous challenges, the benefits of implementing technological advances in the medical sector are monumental. The challenge for the future in this sense is to achieve the full incorporation of technology in clinics, hospitals and laboratories that today are oblivious to it due to the lack of economic resources and with which a significant number of lives would be saved daily, by just as the care and well-being of all citizens would be improved.

2. Motivation

Health is one of the most deficient sectors in Colombia that, in the opinion of many analysts, "is in crisis" [5] and in this context, technology can provide certain benefits that would improve the quality of service of Healthcare Institutions.

Pneumonia accounts for more than 15% of all deaths of children under 5 years of age internationally, in 2015, 920,000 children under 5 years of age died from this disease [6].

Starting from the necessity, Artificial Intelligence (AI) is becoming more common in our lives and becomes a more sophisticated tool that performs the same tasks that humans do, but more efficiently, faster and at a lower costThis is a one of the reasons why including AI in the world of medicine becomes more and more popular in the world [7].

Based on this need, the application of Artificial Intelligence (AI) techniques is becoming more and more common in the health area and is becoming a more sophisticated tool that supports clinical diagnosis more quickly and at a lower cost [8]. This is how the development of an algorithm, supported by Artificial Intelligence, specifically in the use of Neural Networks that allow optimizing the diagnostic processes of Pneumonia, for which CNN Convolutional Neural Networks can help with greater precision through image recognition, the most benefited from this research would be from the doctors who want to carry out research based on it and in turn provide better service and care to patients.

3. Material and Methods

For the development of this work we used the "Team Data Science Process" (TDSP) methodology focused on the life cycle of data science projects.

A significant sample of radiological images of 10,000 images between pneumonia patients and healthy patients was taken for the training of our convolutional neural network, these data were taken from the National Institutes of Health of the United States (National Institute of Health of North America).

The selection criteria was given by the algorithm, the first images of the total set of images, where were mixed those that corresponded to pneumonia and those that did not correspond to pneumonia. The sample taken for the project is 630 training images, 270 images for test and 270 images for final evaluation.

4. Results

4.1. Dataset

This section describes the construction of the data set, detailing step by step how the data set is constructed. In the following flowchart (see Figure 1) the tasks for data acquisition are presented.

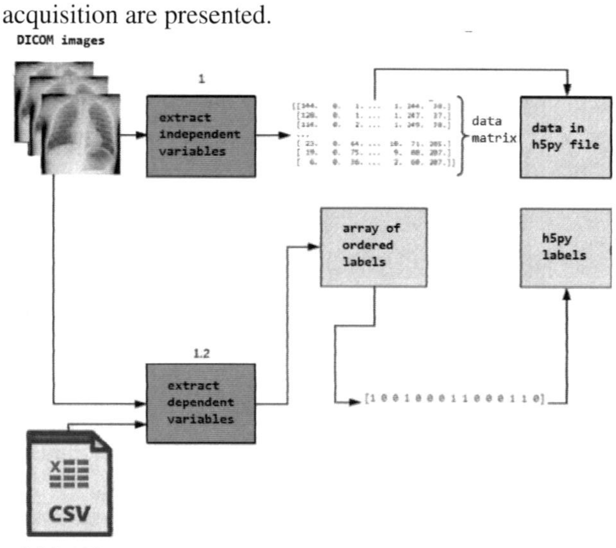

Figure 1Dataset Construction

DICOM (Digital Imaging and Communication in Medicine) images are taken as input data, this DICOM data format contains the information in data sets, that means that a file of a chest X-ray image actually groups the identification of the patient within the file, so that the image can never be separated from this information.
A DICOM data object is constructed from a series of attributes, which correspond to elements such as name, ID, etc. Also a special attribute containing the pixel data of the image, this information provides the algorithm with the necessary data for training and test data set creation. The original name of the DICOM images was previously changed manually to an ascending numerical order, starting from 1 to 10,000, to provide an orderly and sequential reading of each image in the DICOM image data extraction process. For the construction of the dataset, the "*create*_dataset" algorithm was used, in charge of obtaining the data of interest and creating the persistence of these data. In this way it was possible to obtain a given result in organized data, taking into account that the data of interest of each DICOM image is a matrix of pixels/8-bits with values from 0 to 255, which represents an image of size 1024 wide and 1024 high as can be seen in the illustration 7 that in each iteration the images were transformed to a size of 256 high and 256 wide, these images have been read from the local disk and stored in memory by iterations to give it a shape and order (size of the image given an array x number of images) until finally have them serialized in a file of type h5py as shown in Figure 1 . You will be able to understand more specific details in the annex "algorithm for the extraction of dependent and independent variables".

At the same time that the extraction of the image data corresponding to the independent variables was performed, the process of extracting the dependent variables from the "labels_unicos.csv" file, which contains all the diagnoses corresponding to each DICOM image in binary form, was also performed. Each piece of data in the file is in turn associated to each image by means of an identifier (Patientid). These values indicate whether the image has a pneumonia diagnosis or not, which allows the automatic supervised learning algorithm to be built, especially when training and evaluating the CNN (Convolutional Neural Network).

4.2. Modeling CNN

On the other hand, Figure 2 shows the modeling of the convolutional neural network and ends with the evaluations of the constructed model.

IOP Conf. Series: Materials Science and Engineering 1299 (2024) 012002 doi:10.1088/1757-899X/1299/1/012002

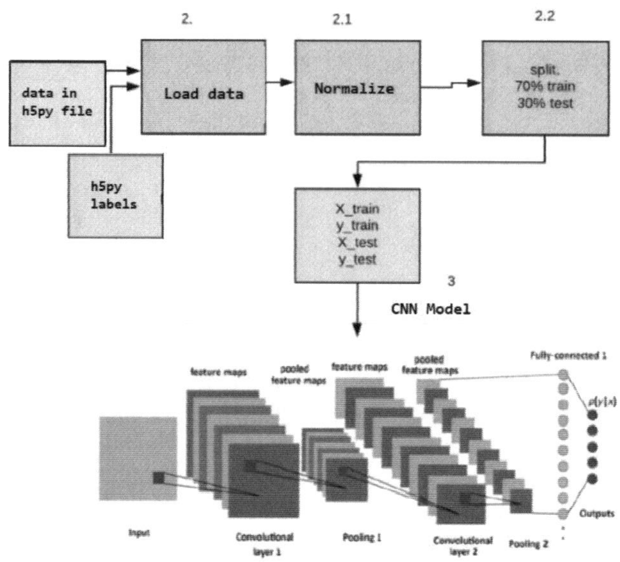

Figure 2 Model construction

The function "load_dataset" it was possible to load in memory the files train.hdf5 (serialized images) and "data_set_Y.hdf5" (serialized labels), taking into account that 900 images were used due to lack of hardware and 270 images were reserved for a final evaluation of the model.
In order to give continuity to the stages:

1. Normalization.
2. Conversion of labels to "one hot encode" format.
3. Split

The normalization stage received the matrix given in columns x rows (900 x 65536) to apply a division between the maximum value of a pixel (255), so that the values stored in the array were light at the time of training and adjustments of the convolutional neural network.
In the conversion of labels to "one hot encode" format, the classes were defined, which for this data set were 2 (non-pneumonia and pneumonia), to obtain a row x column matrix (900 x 2), in which each column represents the class of the image and each row belongs to the result of a specific image, its value in binary was allowing to determine to which category the image belongs in a position as shown in Figure 3 (where [10] = non-pneumonia and [01] = pneumonia).

```
[[1 0]
 [1 0]
 [0 1]
 [1 0]
 [0 1]
 [0 1]
 [0 1]
 [0 1]
 [1 0]
 [0 1]]
```

Figure 3 One hot encode label

The next step is the division of the data. Figure 4 shows the Split technique, which operates with a division of the total data loaded in memory in a distribution of 70% for training data and 30% for test data, the objective of having these data divided is relevant at the moment of evaluating the training with the test data and if the model generalizes with unknown data.

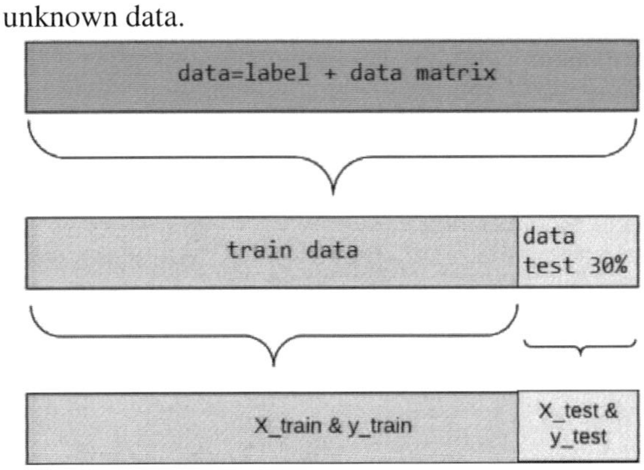

Figure 4 Split data

As a result of the process of dividing 900 "x" data, 630 records are obtained for training and 270 for testing.
For the creation of the CNN (Definition) architecture (See figure 5), a process divided into two sub-processes (feature extraction and classification) was implemented. The method for the construction of a pneumonia classification algorithm consists of two major processes, starting with feature extraction implemented with the CNN technique, which consists of two convolution layers. The second step is a neural network for classification.

IOP Conf. Series: Materials Science and Engineering 1299 (2024) 012002 doi:10.1088/1757-899X/1299/1/012002

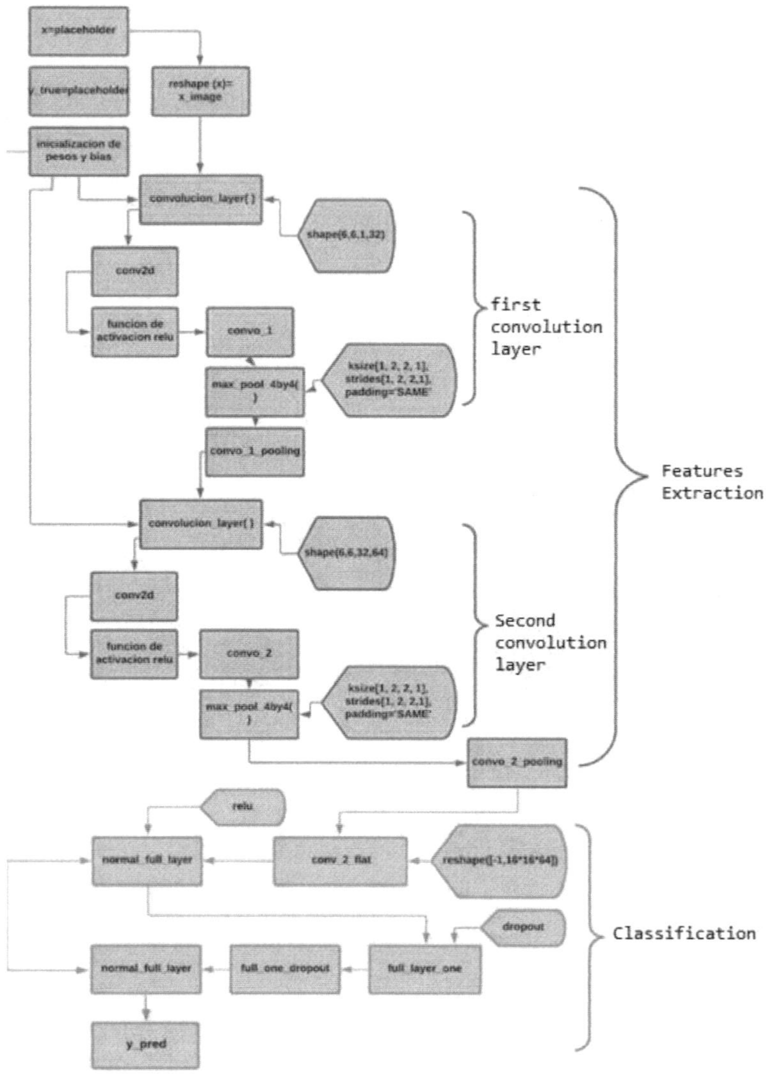

Figure 5 Structure of the model layers

The feature extraction process starts with the application of convolution filters and the application of stride and padding parameters. The filters allow obtaining information from the image in numerical terms, in the form of a feature map and at the end of the convolution process the flattening is performed in the form of a one-dimensional vector (See Figure 6).

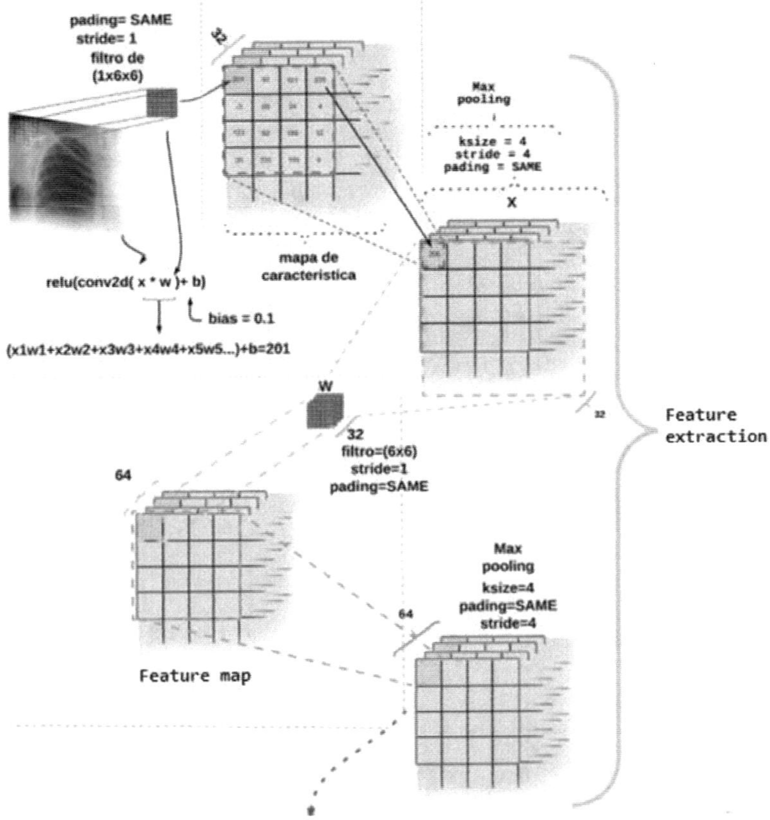

Figure 6 Feature extraction

4.3. Metrics

After training the model, it was possible to observe a decrease in the error for each epoch (number of training interactions) as shown in Figure 7, reaching a value of 0.138335, which represents a good result in terms of error.

```
Cost after epoch 0: 1.804402
Cost after epoch 10: 0.531185
Cost after epoch 20: 0.500114
Cost after epoch 30: 0.448518
Cost after epoch 40: 0.388601
Cost after epoch 50: 0.390130
Cost after epoch 60: 0.329111
Cost after epoch 70: 0.314362
Cost after epoch 80: 0.306148
Cost after epoch 90: 0.246856
Cost after epoch 100: 0.224177
Cost after epoch 110: 0.189994
Cost after epoch 120: 0.187817
Cost after epoch 130: 0.133510
Cost after epoch 140: 0.138335
```

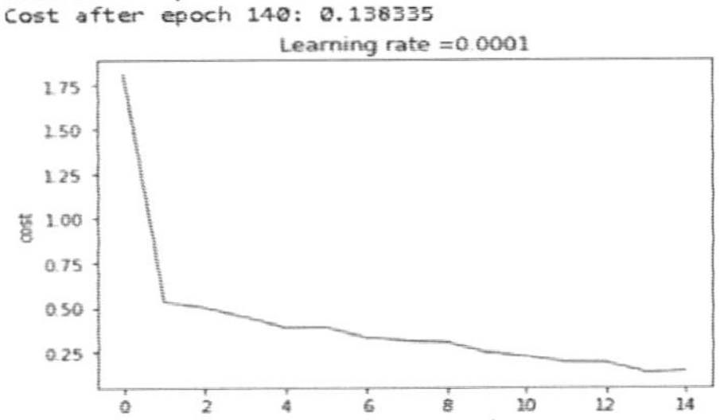

Figure 7 Train process

IOP Conf. Series: Materials Science and Engineering 1299 (2024) 012002 doi:10.1088/1757-899X/1299/1/012002

The results obtained from the model are tested with different parameters and the one with the best performance is selected. Figure 8 shows the record of different parameters (learning rate, batch_size, epochs and steps) that came closest to a significant result, with the report of metrics that were given for each evaluation with the test data. Learning rate (learning rate), Batch size (group of images per epoch), Epoch (iteration in which training and error minimization is performed), Steps (steps taken in each epoch).

	hyperparams				data (30%) & (70%)		Metrics							
	learning rate	batch_size	Epochs	steps per epoch	# data for test	# data for train	Precision	Accuracy	sensitibity	F1	TP	TN	FP	FN
Test	0.0001	32	180	15	270	630	0.77	0.75	0.91	0.83	164	48	42	16
	0.0001	32	100	10	270	630	0.84	0.77	0.81	0.82	146	63	27	34
	0.0002	32	100	10	270	630	0.79	0.76	0.86	0.83	157	50	40	23
	0.0001	32	80	10	270	630	0.80	0.77	0.86	0.83	156	52	38	24
	0.0001	32	100	10	270	630	0.82	0.77	0.83	0.82	150	58	32	30
	0.0001	40	180	10	270	630	0.8	0.72	0.8	0.8	144	54	36	36
	0.0001	32	150	10	270	630	0.78	0.74	0.86	0.82	157	40	40	33
Validacion	0.0001	32	50	10	270	630	0.77	0.74	0.87	0.81	166	31	49	24
	0.0001	32	80	10	270	630	0.79	0.74	82	0.80	156	39	41	34
	0.0001	32	100	10	270	630	0.80	0.72	0.81	0.80	154	43	37	36
	0.0002	32	100	10	270	630	0.83	0.71	0.78	0.81	150	51	29	40
	0.0002	40	100	10	270	630	0.78	0.73	0.80	0.79	153	38	42	37
	0.0002	32	150	10	270	630	0.79	0.72	0.85	0.82	162	39	41	28
	0.0001	32	180	10	270	630	0.75	0.70	0.91	0.82	155	39	41	35
	0.0001	32	150	10	270	630	0.79	0.74	0.82	0.81	157	40	40	33

Figure 8 Test and validation Results

In a given evaluation with the test data there were a total of 270 images which determined the results in a confusion matrix, and 270 normalized images to perform another evaluation test against the results of the test. Figure 9 shows the classification report in terms of precision, accuracy, sensitivity and f1, in comparison with the other results it can be said that these are the highest with the selected parameters.

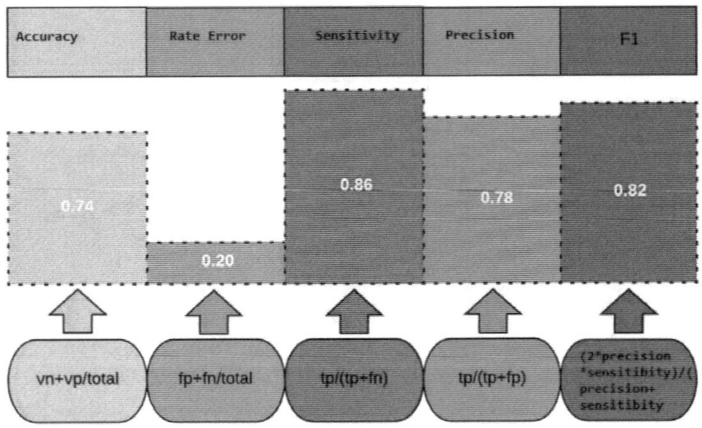

Figure 9 Ranking report with test data

For the evaluation we applied the same parameters that were applied in the test which approached the best possible result with this set of images. The best fit parameters resulting in this confusion matrix (See Figure 10) are: Learning rate = 0.0001, Bacht size = 32, Steps = 12, Epochs = 150.

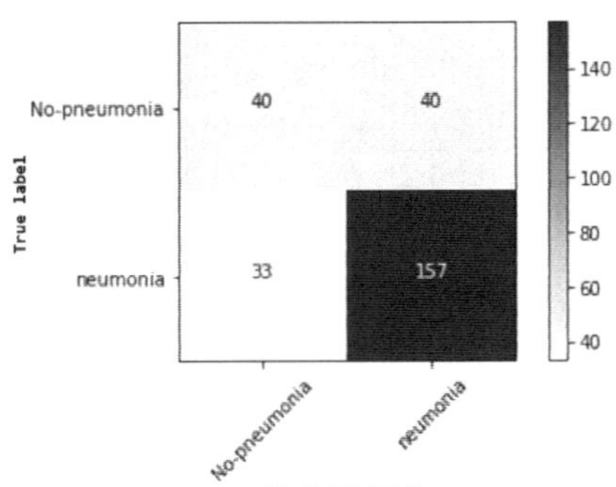

Figure 10 Confusion matrix with evaluation data

5. Conclusions

As a result, the construction of a Machine Learning algorithm was obtained based on an evaluation metric represented in terms of percentage with an accuracy of more than 73% and an error rate of less than 21%. During the course of the research, some limitations arose that were minimized so as not to affect the results so much. For example, the deficiency of computational equipment (hardware) which gave as an observation that it is advisable to acquire computer equipment with RAM memory capacity greater than 56gb, 500gb of storage and any processor of a DSVM (Data Sciencie Virtual Machine) of azure with capacity to work in parallel, given recommendation with the objective to carry out this project to a higher step. If this project scales to a hardware level, it is recommended to analyze the data balancing to decide if it is necessary to implement a data balancing with the "smote" technique of the "imblearn.over_sampling" library.

References

[1] O. M. D. L. S. OMS. [En línea]. Available: https://www.who.int/mediacentre/news/releases/2014/child_mortality_estimates/es /. [Último acceso: 15 MARZO 2019].

[2] DANE, «DIEZ PRIMERAS CAUSAS DE DEFUNCIÓN, SEGÚN DEPARTAMENTO DE RESIDENCIA,» Santa Fe de Bogotá, 2016.

[3] H. C. U. d. S. d. Compostela, «NEUMO EXPERTOS EN PREVENCION,» Hospital Clínico Universitario de Santiago de Compostela, [En línea]. Available: https://neumoexpertos.org/2017/05/17/pronostico-y-mortalidad-de-la-neumonia/.

[4] OMS, «https://www.who.int/es/news-room/fact-sheets/detail/pneumonia,» 07 11 2016. [En línea]. Available: https://www.who.int/es/news-room/fact- sheets/detail/pneumonia. [Último acceso: 5 11 2018].

[5] E. TIEMPO, «EL TIEMPO - BLOGS,» ELTIEMPO.COM, [En línea]. Available: http://blogs.eltiempo.com/motor-de-busqueda/2017/03/05/la-crisis-del-sistema-de- salud/.

[6] S. R. d. A. d. Norte, «kaggle.com,» Sociedad Radiologica de America del Norte, 2018. [En línea]. Available: https://www.kaggle.com/c/rsna-pneumonia-detection-challenge. [Último acceso: 15 enero 2019].

[7] clustersalud.americaeconomia.com, «americaeconomia.com,» clustersalud.americaeconomia.com, 2018. [En línea]. Available: https://clustersalud.americaeconomia.com/insumos-y-servicios-hospitalarios/5- problemas-del-sectorsalud-que-la-inteligencia-artificial-puede-resolver. [Último acceso: 13 12 2018].

[8]«https://ingenieriaelectronica.org/precision-exactitud-y-sensibilidad-mediciones-electricas/,» 2018. [En línea]. [10] C. d. D. y. D. d. I. Informática, «coddii.org,» La opinión de Murcia, [En línea].

Analysis of the influence of FDM parameters in the tensile strength response using machine learning

V Cudris [1] [1], V J Santiago [2], C A Londoño [1] and H Lopez [1]

[1] Mechatronics Engineering Program, Institucion Universitaria ITSA, Soledad-Atlantico, COL
[2] Industrial Engineering Program, Universidad Libre, Barranquilla, COL

E-mail: vcudris@itsa.edu.co

E-mail: veraj.santiagom@unilibre.edu.com

E-mail: calondono@itsa.edu.co

E-mail: hlopez@itsa.edu.co

Abstract. This research presents a comprehensive experimental study on the effect of temperature, material and process parameters related to tensile strength response in 3D printing manufacturing process with ABS material. A Hyper Latin Square design was chosen for the experimental points distribution. Thirteen parameters with multiple levels are considered LAYERHEIGHT, WALLTHICKNESS, TOPBOTTOMTHICKNESS, TOPBOTTOMLINEDIRECTION1, TOPBOTTOMLINEDIRECTION2, INFILLDENSITY, INFILLLINEDIRECTION1, INFILLLINEDIRECTION2, PRINTSPEED, EXTRUSIONTEMP, BEDTEMP, WORKSPACETEMP and POSITION. Type IV tensile specimens are fabricated and tested with an universal testing machine. Maximum stress is measured, evaluated and analyzed in three different building positions.Machine learning algorithm with Orange Data mining software are used to study underlying relations between factors and response. Experimental results indicate that INFILLDENSITY, TOPBOTTOMTHICKNESS and INFILLLINEDIRECTION1 has a strong positive correlation with tensile strength. Meanwhile, TOPBOTTOMLINEDIRECTION1, WORKSPACETEMPERATURE and PRINTSPEED has a negative correlation with tensile strength. Position 1 with line depositions parallel to Y axis produce the higher tensile strength response. Findings imply that machine algorithms can be used to study multiple parameters at time.

Keywords: fused deposition modelling, print 3D, parameters, machine learning, tensile strenght

1. Introduction

Additive manufacturing (AM) is a component manufacturing process by modeling in computer-aided design (CAD) software and then mechanically bonding materials, which are layered one on top of the other. There are different types of MA, one of them is fused deposition modeling (FDM), patented in 1989 [1].

The most popular method of FDM is 3D printing, in which various process parameters must be controlled, such as process parameters (deposition angle, layer height, fabrication direction, deposition width, print speed, density of fill, air gaps, fill pattern, extrusion temperature),

[1] Present address: Mechatronics Engineering Program, Institucion Universitaria ITSA, Soledad-Atlantico, COL

workspace parameters (bed temperature, chamber temperature, humidity, nozzle diameter) and material parameters (material type, color, diameter of the filament) [2].

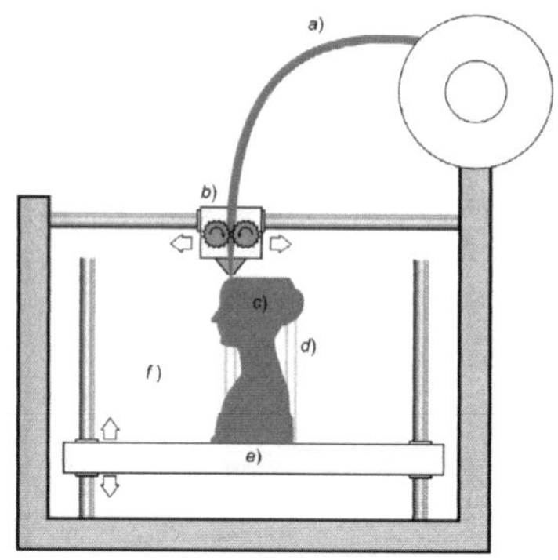

Figure 1. Schematic representation of Print 3D basic machine. a)Filament, b)Extruder, c)Model, d)Supports, e)Bed, f)Work space

Regarding 3D printing, the parameters and their influence on the properties and behavior of the manufactured part have been investigated. Much of this research is oriented toward mechanical properties such as resistance to tension, compression, bending, torsion and impact energy absorption.[1] [2] [3].

Tensile stress has been extensively studied by various authors, and is one of the most important properties of parts manufactured by 3D printing, whose application must withstand this type of stress to a great extent. These studies have been carried out using different statistical techniques such as factorial design [4, 5, 6, 7, 8, 9], response surface methodology [10, 11, 12, 13, 14], Taguchi [15, 16, 17, 18, 19] and machine learning algorithms [20] [21].

2. Methodology

2.1. 3D print Material

The material used for the study was acrylonitrile butadiene styrene (ABS). It is the most commonly used petroleum-derived plastic in 3D printing. It is characterized by its high mechanical and thermal resistance and its great flexibility. The ABS is from the eSUN brand, red color, with the properties and printing parameters recommended in Table 1.

2.2. 3D printing machine

The printer used for the fabrication of the test specimens is a INTAMSYS FUNMAT HT (Figure 2) with a build volume of 260 x 260 x 260 mm with 50-micron high-resolution industrial quality 3D printing. The thermal system design of this 3D printer includes a 90 °C constant temperature chamber, 160 °C heated build plate and 450 °C high-temperature extruder with an all-metal hot end.

2.3. Printing process parameters

Ten process parameters are included for the research:

- Temperature parameters
  - Extrusion temperature (EXTRUSIONTEMP)
  - Platform temperature (BEDTEMP)
  - Work space temperature (WORKSPACETEMP)

Table 1. eSun ABS properties and recommended printing parameters

3D PRINTING FILAMENT	ABS
Density(g/cm3)	1.04
Heat Distortion Temp(C,0.45MPa)	78
Melt Flow Index(g/10min)	12(220C/10kg
Tensile Strength(MPa)	43
Elongation at Break(%)	22
Flexural Strength(MPa)	66
Flexural Modulus(MPa)	1177
IZOD Impact Strength(kJ/m^2)	29
Extruder Temperature( )	230 – 270 Recommended temperature240
Bed temperature()	95 – 1110 °C
Fan Speed	0%
Printing Speed	40 – 100mm/s
Heated Bed	Required

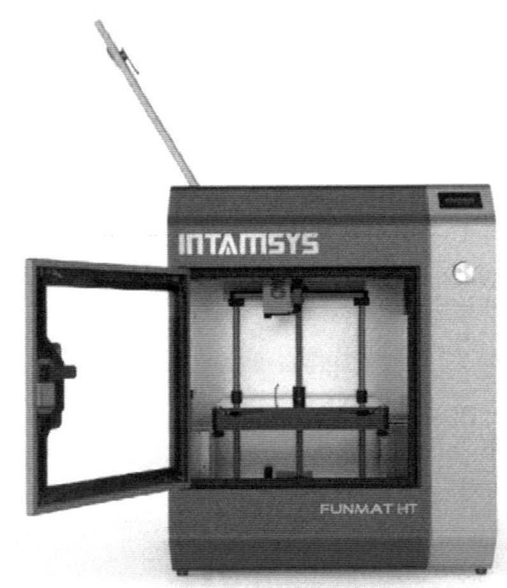

Figure 2. INTAMSYS 3d printer (https://intamsys.wpenginepowered.com/wp-content/uploads/2021/07/Funmat-ht-1024px-2.jpg)

- Machine parameters
  - Print velocity (PRINTSPEED)
- Material deposition parameters
  - Layer height (LAYERHEIGHT)
  - Wall thickness (WALLTHICKNESS)
  - Top and bottom layer thickness (TOPBOTTOMTHICKNESS)
  - Top and bottom layer line directions (TOPBOTTOMLINEDIRECTION)
  - Infill density (INFILLDENSITY)
  - Infill line direction (INFILLLINEDIRECTION)
  - Build Orientation (POSITION)

2.4. Factor parameter levels

A Latin Hypercube design was chosen to assess the impact of the process factors on the output. This experimental design offers a more uniform spreading of the random sample experimental points [22]. Level ranges were established base on previous researches and are show in Table 2

Table 2. Factors an range for the experimental points sample design

FACTOR	LEVEL INTERVAL
(LH) LAYERHEIGHT (mm)	0.1, 0.2, 0.3
(WT) WALLTHICKNESS	0.6, 0.8, 1
(TBT) TOPBOTTOMTHICKNESS(mm)	0.4, 0.8, 1
(TBLD1) TOPBOTTOMLINEDIRECTION1(°)	0, 180
(TBLD2) TOPBOTTOMLINEDIRECTION2(°)	0, 180
(ID) INFILLDENSITY(%)	20, 50, 80, 100
(IDLD1) INFILLLINEDIRECTION1(°)	0, 180
(ILD2) INFILLLINEDIRECTION2(°)	0, 180
(PS) PRINTSPEED(mm/sec)	40, 60, 80
(ET) EXTRUSIONTEMP(°C)	220, 260
(BT) BEDTEMP(°C)	80, 100
(WST) WORKSPACETEMP(°C)	30, 60
(POS) POSITION	P1, P2, P3

Applying the LHS for the distribution of the experimental points, a set of data is obtained as shown in Table 3

2.5. Printing process

Models from SolidWorks were saved as .stl type and preprocessed in Ultimaker Cura slicer software. Tests specimens were printed in the three positions with the same parameters to compare the effect of the same experimental point in the three building positions, as shown in Figure 3.

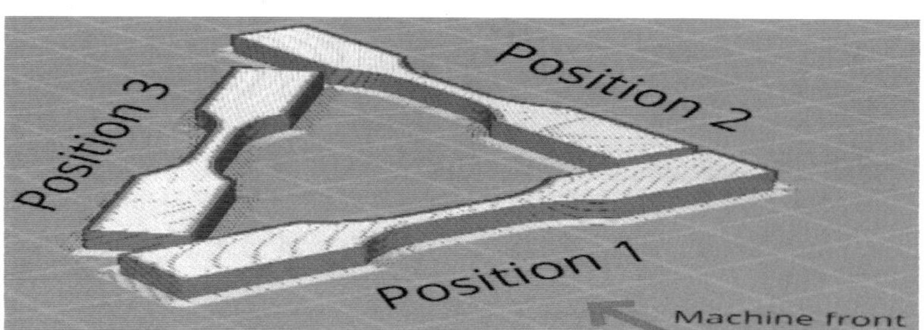

Figure 3. The three positions of the testing specimens

2.6. Tensile testing

2.6.1. Testing specimen Specimen for the tension test is a ASTM D638 Type IV [23]. The specimen was modeled in SolidWorks software with dimensions as in Figure 4.

The Universal testing machine is a 50 kN Armsfield with a ±0.5% precision. Test's velocity was set in $1\frac{mm}{sec}$.

Table 3. Experimental points generate with LHS

RUN	LH	WT	TBT	TBLD1	TBLD2	ID	ILD1	ILD2	WST	ET	BT	PS	POS
1	0.21	0.78	0.87	140	80	64.44	60	20	46.67	220.00	100.00	44.44	P1
2	0.21	0.78	0.87	140	80	64.44	60	20	46.67	220.00	100.00	44.44	P2
3	0.10	0.96	0.80	80	40	91.11	180	160	36.67	260.00	91.11	62.22	P1
4	0.10	0.96	0.80	80	40	91.11	180	160	36.67	260.00	91.11	62.22	P2
5	0.10	0.96	0.80	80	40	91.11	180	160	36.67	260.00	91.11	62.22	P3
6	0.30	0.82	0.73	0	0	28.89	20	40	33.33	224.44	97.78	66.67	P1
7	0.30	0.82	0.73	0	0	28.89	20	40	33.33	224.44	97.78	66.67	P2
8	0.30	0.82	0.73	0	0	28.89	20	40	33.33	224.44	97.78	66.67	P3
9	0.19	1.00	1.00	60	120	55.56	120	0	30.00	255.56	93.33	40.00	P1
10	0.19	1.00	1.00	60	120	55.56	120	0	30.00	255.56	93.33	40.00	P2
11	0.19	1.00	1.00	60	120	55.56	120	0	30.00	255.56	93.33	40.00	P3
12	0.12	0.60	0.53	160	140	46.67	140	100	50.00	228.89	95.56	57.78	P1
13	0.12	0.60	0.53	160	140	46.67	140	100	50.00	228.89	95.56	57.78	P2
14	0.12	0.60	0.53	160	140	46.67	140	100	50.00	228.89	95.56	57.78	P3
15	0.28	0.64	0.93	20	180	100.00	160	120	40.00	233.33	86.67	48.89	P1
16	0.28	0.64	0.93	20	180	100.00	160	120	40.00	233.33	86.67	48.89	P2
17	0.28	0.64	0.93	20	180	100.00	160	120	40.00	233.33	86.67	48.89	P3
18	0.14	0.69	0.40	100	20	37.78	0	80	56.67	242.22	84.44	75.56	P1
19	0.14	0.69	0.40	100	20	37.78	0	80	56.67	242.22	84.44	75.56	P2
20	0.14	0.69	0.40	100	20	37.78	0	80	56.67	242.22	84.44	75.56	P3
21	0.17	0.87	0.47	40	160	82.22	100	140	60.00	246.67	82.22	80.00	P1
22	0.17	0.87	0.47	40	160	82.22	100	140	60.00	246.67	82.22	80.00	P2
23	0.17	0.87	0.47	40	160	82.22	100	140	60.00	246.67	82.22	80.00	P3
24	0.23	0.91	0.60	120	100	20.00	80	180	43.33	251.11	88.89	53.33	P1
25	0.23	0.91	0.60	120	100	20.00	80	180	43.33	251.11	88.89	53.33	P2
26	0.23	0.91	0.60	120	100	20.00	80	180	43.33	251.11	88.89	53.33	P3
27	0.26	0.73	0.67	180	60	73.33	40	60	53.33	237.78	80.00	71.11	P1
28	0.26	0.73	0.67	180	60	73.33	40	60	53.33	237.78	80.00	71.11	P2
29	0.26	0.73	0.67	180	60	73.33	40	60	53.33	237.78	80.00	71.11	P3

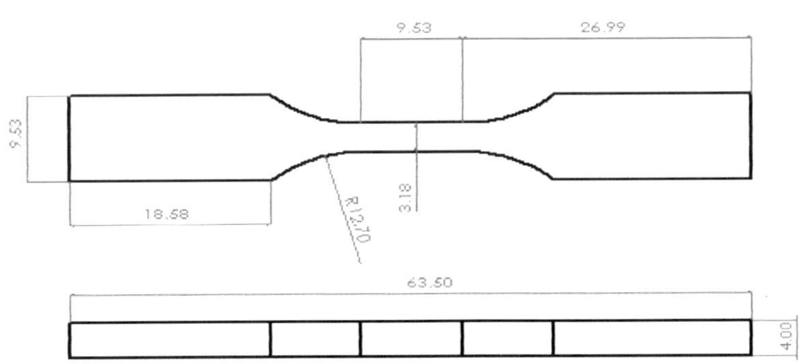

Figure 4. ASTM D638 Type IV Tension test specimen (dimensions in mm)

After the test, stress-strain curves and results were registered in the machine software. Yield, ultimate and break stress, and maximum load were measured from the graphics. Young's modulus was calculated from the slope of the stress-strain curves.

IOP Conf. Series: Materials Science and Engineering 1299 (2024) 012003 doi:10.1088/1757-899X/1299/1/012003

2.7. Machine learning processing

For machine learning analysis, a visual data mining tool was applied. Orange Data Mining [24] is a computer program for data mining and predictive analytics developed at the Faculty of Informatics at the University of Ljubljana. It consists of a series of components developed in C++ that implement data mining algorithms, as well as data preprocessing and graphical representation operations as in Figure 5. The principal advantage is that to explore data with Orange requires no programming or in-depth mathematical knowledge.

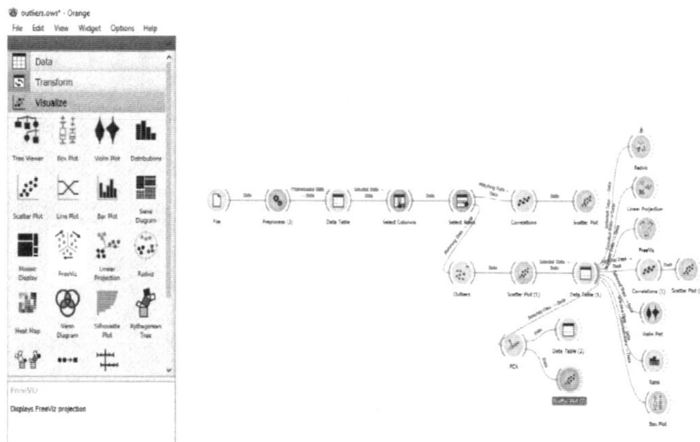

Figure 5. Orange workflow example

3. Results and Analysis

3.1. Tensile test results

Table 4 show calculated results for the tension tests.

Table 4. Tension test results

RUN	TIEMPO (min)	MAX LOAD (N)	BREAK LOAD (N)	MAX STRESS (MPa)	DEF MAX	BREAK STRESS (MPa)	DEF BREAK	E (MPa)
1	46	255	200	21.15	0.214	16.587	0.258	159.11
2	46	267	217	22.14	0.614	19.41	0.648	147.28
3	64	435	416	36.08	0.47	34.5	0.488	258.71
4	64	255	231	21.15	0.256	19.16	0.28	149.55
5	64	323	292	26.79	0.442	24.22	0.47	167.3
6	20	128	112	10.62	0.172	9.29	0.17	90.23
7	20	179	150	14.84	0.204	12.44	0.216	101.26
8	20	188	166	15.6	0.28	13.77	0.28	83.23
9	54	232	173	19.24	0.204	15.67	0.234	112.79
10	54	248	186	20.57	0.294	15.43	0.37	114.14
11	54	212	189	17.58	0.252	15.67	0.268	151.02
12	47	138	123	11.4	0.184	10.2	0.189	89.5
13	47	123	121	10.2	0.238	10.03	0.24	70.11
14	47	94	82	7.79	0.21	6.8	0.218	98.1
15	40	300	236	24.88	0.2	19.57	0.254	191.04
16	40	318	262	26.37	0.178	21.73	0.254	218.87
17	40	289	267	24	0.134	22.14	0.16	223.82
18	28	153	136	12.69	0.256	11.28	0.266	107.57
19	28	157	137	13.02	0.234	11.36	0.254	106.57
20	28	142	97	11.77	0.298	8.04	0.374	84.71
21	30	335	322	27.78	0.468	26.7	0.476	224.92
22	30	210	1.58	14.42	0.214	13.1	0.264	145.04
23	30	192	168	15.92	0.144	13.93	0.158	148.13
24	27	178	92	14.76	0.23	7.63	0.314	109.17
25	27	160	135	13.27	0.284	11.2	0.298	75.3
26	27	153	120	12.69	0.284	9.95	0.322	70.04
27	23	206	174	17.08	0.218	14.43	0.254	104.11
28	23	176	128	14.6	0.2	10.62	0.234	131.54
29	23	185	172	15.34	0.174	14.26	0.198	137.87

IOP Conf. Series: Materials Science and Engineering 1299 (2024) 012003 doi:10.1088/1757-899X/1299/1/012003

3.2. Outliers

Outliers were calculated with local outlier based on the local density from the k-nearest neighbors factor method [25]. Three outliers were found, run 6, 7 and 8, as shown in Figure 6.

Hereinafter, next results are analyzed with the inliers points, ignoring outliers.

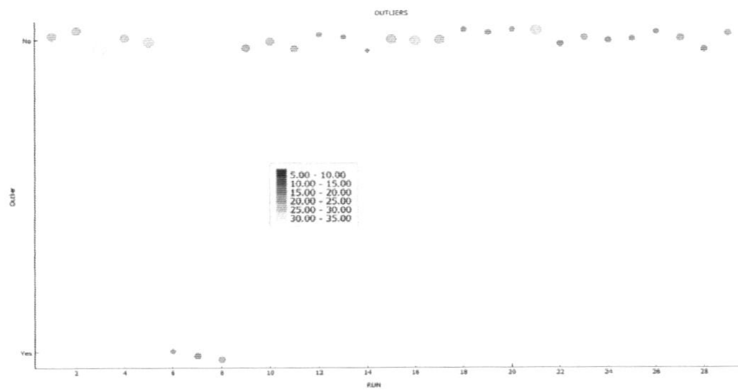

Figure 6. Outliers widget

3.3. Correlation results

For the correlation's analysis, Pearson method was computed[26]. Results in Table 5 shows that positive and negative correlation exist between factors and the maximum stress.INFILLDENSITY has the biggest correlation coefficient among the positive correlations, followed by TOPBOTTOMTHICKNESS and INFILLLINEDIRECTION1.

TOPPBOTTOMLINEDIRECTION1 has the biggest negative correlation factor, with WORKSPACETEMPERATURE and PRINTSPEED as the only three negative.

INFILLDENSITY, TOPBOTTOMTHICKNESS and INFILLLINEDIRECTION1 shows a strong positive correlation (¿0.5), meanwhile WALLTHICKNESS, EXTRUSSIONTEMPERATURE and INFILLDIRECTION2 swows weak positive correlation. LAYERHEIGHT, TOPBOTTOMLINEDORECTION2 and BEDTEMPERATURE has a correlation closes to zero (weak correlation).

Table 5. Correlation coefficients with respect to maximum stress

+0.712	INFILLDENSITY	MAX STRESS
+0.584	MAX STRESS	TOPBOTTOMTHICKNESS
-0.534	MAX STRESS	TOPBOTTOMLINEDIRECTION1
+0.510	INFILLLINEDIRECTION1	MAX STRESS
-0.476	MAX STRESS	RUN
-0.436	MAX STRESS	WORKSPACETEM
+0.343	MAX STRESS	WALLTHICKNESS
+0.245	EXTRUSIONTEMP	MAX STRESS
-0.206	MAX STRESS	PRINTSPEED
+0.106	INFILLLINEDIRECTION2	MAX STRESS
+0.081	LAYERHEIGTH	MAX STRESS
+0.077	MAX STRESS	TOPBOTTOMLINEDIRECTION2
+0.061	BEDTEMP	MAX STRESS

IOP Conf. Series: Materials Science and Engineering 1299 (2024) 012003 doi:10.1088/1757-899X/1299/1/012003

3.4. Visual Explorative data analysis

Three visualization techniques for explorative data analysis, Linear Projection, RadViz y FreeViz were used to analyze relations between factors and the tension stress response, and compare with correlation results. Visual tools are show in Figures 7, 8 and 9. The trend of the positive and negative factors is the same as in Table 5, with INFILLDENSITY, TOPBOTTOMTHICKNESS and INFILLLINEDIRECTION1 in direct strong correlation with the response.

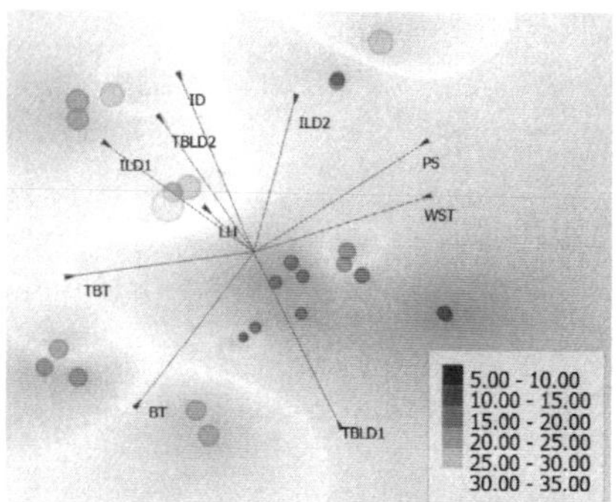

Figure 7. Linear projection

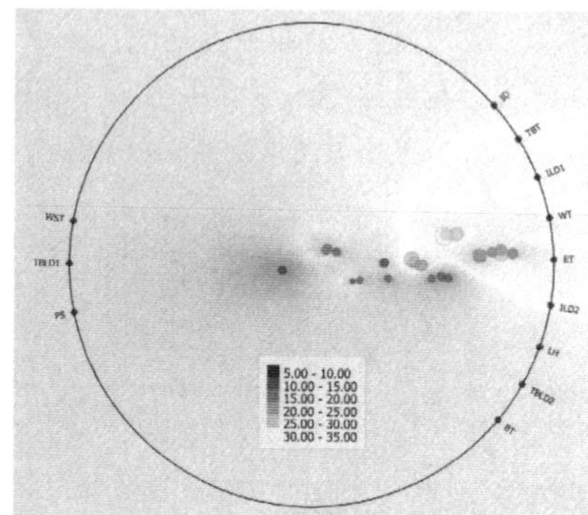

Figure 8. RadViz

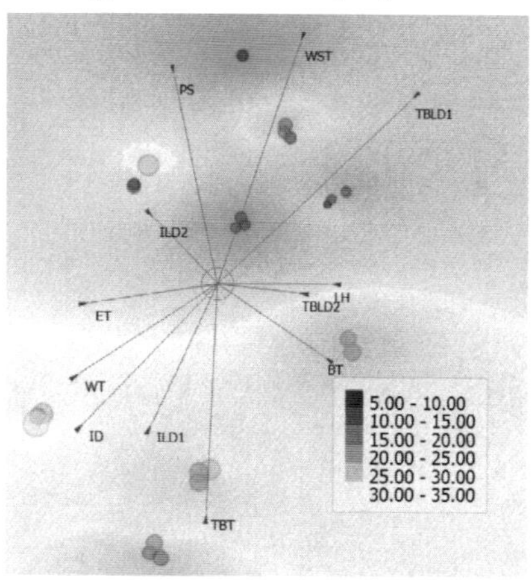

Figure 9. FreeViz.

3.5. Violin and Box plots for Printing positions

Most of the ML algorithms work with numerical factors or categorical factors with only two levels. Position factor has three levels, in this case violin and box plot are used for the visual comparison of the results in Figure 10 and 11.

Printing Position 1 has the biggest mean and the higher value in respect to tensile stress response. A value of 20.56 MPa against 17.30MPa and 16.48MPa for position 2 and 3.

3.6. Classification tree algorithm

Classification tree algorithm works well with numerical and categorical variables. In this case, we can use decision tree to analyze numerical variables with the printing position.

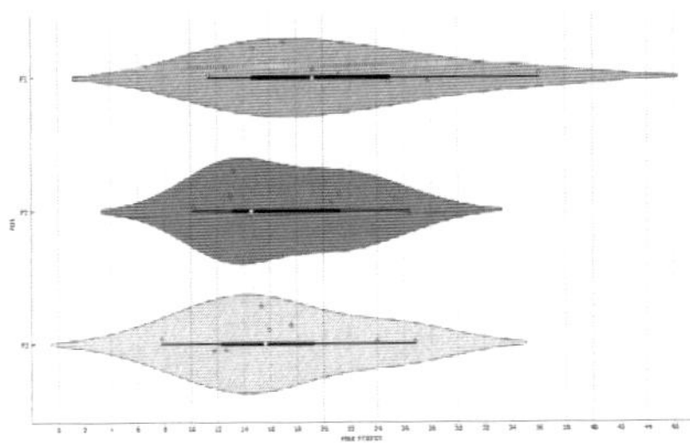

Figure 10. Violin plot for POS Vs. maximum stress.

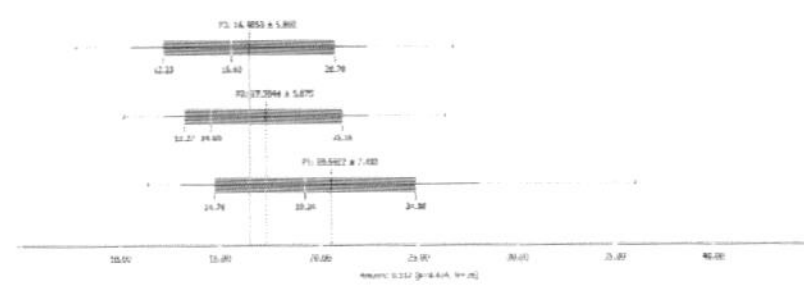

Figure 11. Box plot for POS Vs. maximum stress.

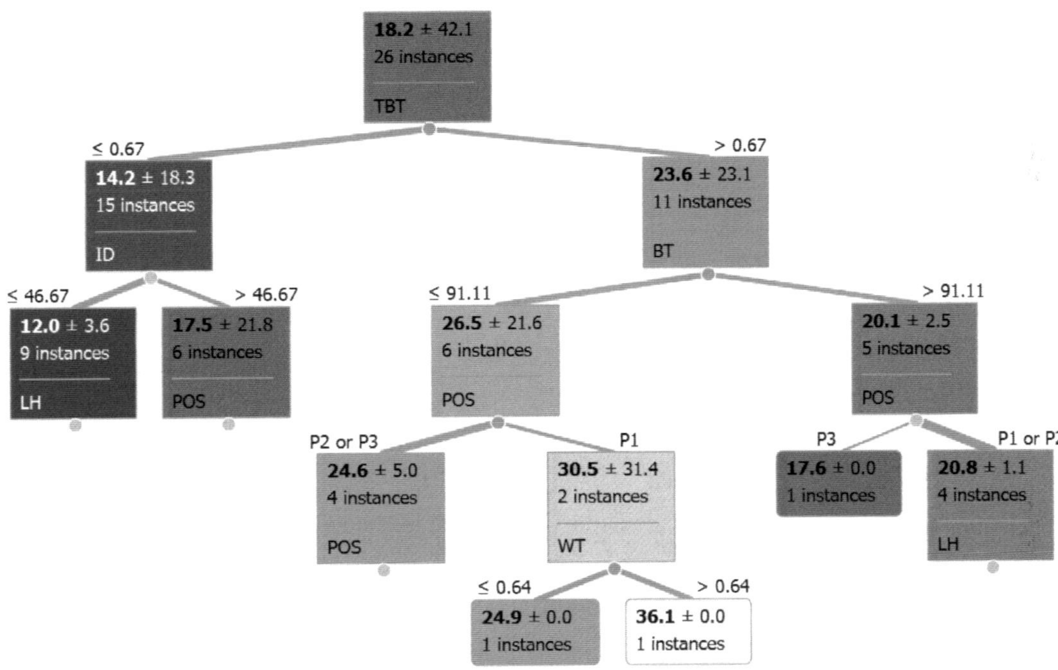

Figure 12. Classification tree

Figure 12 shows the interaction between numerical and categorical factors. The higher response for tensile strength is obtained with a TOPBOTTOMTHICKNESS > 0.97 mm, BEDTEMPERATURE < 91.11 °C, POSITION = 1, WALLTHICKNESS > 0.64 mm.

4. Conclusions and Future Works

Fourteen parameters were analyzed using machine learning techniques for a 3D printing precess. Significant correlation were identified, with strong and weak relations between parameters and the tensile strength response.

It was found that INFILLDENSITY, TOPBOTTOMTHICKNESS and INFILLLINEDIRECTION1 increase the tensile strength response.

To a lesser extent, the factors WALLTHICKNESS, EXTRUSIONTEMP and INFILLLINEDIRECTION2 produce an increase in the response.

IOP Conf. Series: Materials Science and Engineering 1299 (2024) 012003 doi:10.1088/1757-899X/1299/1/012003

TOPBOTTOMLINEDIRECTION1, WORKSPACETEMP and PRINTSPEED affects in negative manner, the tensile strength responseonse with a significant influence.

Printing position 1 produce the best tensile strength response compared with the other two positions.

For future works, research on the deposited line directions in the infill and top-bottom with diffeferent infill patterns shall be considered.

References

[1] Vyavahare S, Teraiya S, Panghal D and Kumar S 2019 *Rapid Prototyping Journal*
[2] Gao G, Xu F, Xu J, Tang G and Liu Z 2022 *Micromachines* **13** ISSN 2072-666X URL `https://www.mdpi.com/2072-666X/13/4/553`
[3] Dey A and Yodo N 2019 *Journal of Manufacturing and Materials Processing* **3** 64
[4] Rosid I A and Tontowi A E 2021 *OPSI*
[5] Haidiezul A H M, Haidiezul A H M, Haidiezul A H M, Hazwan M H M, Hazwan M H M, Hazwan M H M, Hazwan M H M, Lee W S, Gunalan, Najihah N F and Fadhli I 2020 *IOP Conference Series: Materials Science and Engineering*
[6] Bialete E R, Manuel M C E, Alcance R M E, Canlas J P A, Chico T J B, Sanqui J P, Cruz J C D and Verdadero M S 2020 *2020 IEEE 12th International Conference on Humanoid, Nanotechnology, Information Technology, Communication and Control, Environment, and Management (HNICEM)*
[7] Rodríguez P, Zapico P, Robles P E, Soto A and Barreiro J 2021 *IOP Conference Series: Materials Science and Engineering*
[8] Pawar S, Dolas D and Dolas D R 2020 *null*
[9] Nancharaiah T and Nancharaiah T 2011 *null*
[10] Srinivasan R, Srinivasan R, Srinivasan R, Srinivasan R, Pridhar T, Pridhar T, Ramprasath L, Charan N S and Ruban W 2020 *Materials Today: Proceedings*
[11] Gao G, Xu F and Xu J 2022 *Machines*
[12] Mohamed O A, Masood S H and Bhowmik J L 2016 *Applied Mathematical Modelling*
[13] Waseem M, Waseem M, Salah B, Habib T, Saleem W, Abas M, Khan R, Ghani U and Siddiqi M U R 2020 *Polymers*
[14] Vorkapić M, Mladenovic I, Pergal M V, Ivanov T and Baltić M 2022 *Tribology and Materials*
[15] Hikmat M, Rostam S, Rostam S, Ahmed Y M and Ahmed Y M 2021 *null*
[16] Kam M, Ipekçi A, Ömer Şengül and Ömer Şengül 2021 *Journal of Thermoplastic Composite Materials*
[17] Maguluri N, Suresh G and Rao K V 2021 *Journal of Thermoplastic Composite Materials*
[18] Sharma K, Kumar K, Singh K R and Rawat M 2021 *IOP Conference Series: Materials Science and Engineering*
[19] Abouelmajd M, Bahlaoui A, Arroub I, Zemzami M, Hmina N, Lagache M and Belhouideg S 2021 *International Journal for Simulation and Multidisciplinary Design Optimization*
[20] Goh G D, Sing S L and Yeong W Y 2021 *Artificial Intelligence Review* **54** 63–94
[21] Johnson N, Vulimiri P, To A, Zhang X, Brice C, Kappes B and Stebner A 2020 *arXiv preprint arXiv:2005.05235*
[22] Tirthajyoti Tirthajyoti/doepy: Design of experiment generator. read the docs at: Https://doepy.readthedocs.io/en/latest/ URL `https://github.com/tirthajyoti/doepy`
[23] ASTM 2014 Astmd638 - standard test method for tensile properties of plastics URL `https://www.astm.org/d0638-14.html`
[24] Ljubljana Bioinformatics Laboratory U o Orange data mining URL `https://orangedatamining.com/`
[25] Ljubljana Bioinformatics Laboratory U o Outliers URL `https://orangedatamining.com/widget-catalog/unsupervised/outliers/`
[26] Ljubljana Bioinformatics Laboratory U o Correlations URL `https://orangedatamining.com/widget-catalog/unsupervised/correlations/`

IOP Conf. Series: Materials Science and Engineering 1299 (2024) 012004 doi:10.1088/1757-899X/1299/1/012004

Optimization of the diesel photovoltaic isolated system installed in the Nazareth community using wind energy.

L M Perpiñán[1,4], M J Bastidas[2], S D Vergara[1] and V J Iguarán[3]

[1] Department of Mechanical Engineering, Faculty of Engineering, University of Guajira, Kilometro 5 vía Maicao, Colombia, Investigation Group DESTACAR.
lperpinan@uniguajira.edu.do
[2] Department of Mechanical Engineering, Faculty of Engineering, University of Guajira, Kilometro 5 vía Maicao, Colombia, director of the research group DESTACAR
[3] Faculty of Basic and Applied Sciences, Biology, University of Guajira, Kilometro 5 vía Maicao, Colombia.
[4] Author to whom any correspondence should be addressed.

Abstract. The objective of the work is the optimization of the isolated and hybridized electric power generation system with photovoltaic technology and genset, installed in the village of Nazareth belonging to the municipality of Uribia, Colombia. The optimization criteria established in this study consists of respecting what is already installed and from there propose various search spaces in HOMER Energy to find the optimal system configuration, the main idea is to minimize the consumption of diesel fuel through the inclusion of a wind system. two optimal scenarios are proposed, the first minimizes the participation of the genset to 10%, the second proposes a system of 100% renewable energy generation. Based on the results of the study, it is concluded that the system proposed in scenario C is the most pertinent alternative, It is concluded that the system proposed in scenario C is the most relevant alternative, because the system has 90% autonomy without the generator set.

1. Introduction

Renewable energies have currently been supported worldwide by many countries, cataloging them as a means of generating healthy energy for the planet [1]. The installation of this type of technology is very flexible because it can be adapted to local conditions and is used mostly to supply power to remote places with difficult access [2]. The consumption of renewable energy in the world has been increasing by 2.3% since 2015, which means that global carbon emissions associated with renewable energy remained stable in 2014, while the world economy grew [3], such favorable effects are attributed to the use of renewable energies and improvements in energy efficiency.

According to the US Energy Information Administration (EIA) International Energy Outlook 2013, the current world demand for primary energy is 546.8 quadrillion Btu (546.8 × 10 15 Btu) and is projected to reach to nearly 820 quadrillion Btu by 2040. Global energy consumption of electrical power is estimated to double in the next 15 to 20 years. China, India, Morocco, and Chile, among other countries, are projected to increase their electricity consumption in the next two decades [4].

Part of the Colombian national territory is not covered by the interconnected energy system. Most of the Non-Interconnected Zones (ZIN) meet their energy needs through Diesel generators, which implies a high cost in fuel transportation and equipment maintenance, the energy supply in vulnerable communities is only available approximately 4 hours a day [5]. As pointed out by the 2012 global energy competitiveness study, conducted by the Choiseul institute and Kpmg. "Power generation in Colombia is one of the pillars of economic development" [6]. Colombia is a country with great natural wealth such as hydro potential, gas and coal, it also has wind and solar resources and the use of biomass, the abundance of natural resources is evident. Some studies carried out reveal the different areas of the country where there is a great variety of renewable resources, which can be used for energy production [7].

The energy demand has registered an increase since 2001 with rates that oscillate between 1.5% and 4.1% per year, the national territory has a multiannual daily solar average close to 4.5 kWh/m2, where the peninsula of La Guajira stands out with a average of 6.0 kWh/m2 and wind speeds equal to 5 m/s reaching 11m/s [8]. Although Colombia has energy wealth, currently it only has one wind farm installed; Jepirachi Park, located near Cabo de la Vela in La Guajira. Public Companies of Medellín (EPM) was in charge of its installation and started up since April 2004 [9]. According to a report delivered by EPM, the wind speedsof the place are favorable almost all year round, isolated winds coming from the east-west with low intensityof turbulence. The park is made up of 15 wind turbines manufactured by NORDEX, a German company, each with 1.3 MW of power, which is equivalent to 19.5 MW of total power [9].

Years ago, the community of Nazareth had a generator power generator, in 2011, the company ADES proposed to the Institute for Planning and Promotion of Energy Solutions for Non-Interconnected Areas (IPSE) a consistent solution to minimize the operation of the generator set, through the incorporation of renewable energy sources (wind and solar) in the existing system. The proposal proposed incorporating two 100 kW ADES pivoting single-blade wind turbines at a height of 20 m, a 100kW photovoltaic park with tracking, a battery bank and three generator sets totaling 250 kW of power. The system was installed, according to the local press and technical reports from the IPSE, the wind technology presented problems in its operation, currently only the photovoltaic system and the Diesel generator are operating [10].

La Guajira is a beautiful department with great natural wealth, it has great marine wealth and resources from nature such as mineral coal, natural gas, gypsum, barite, sea salt, gold, marble, clay, and limestone among others. Geographically it is the northernmost territory of Colombia and South America. It is characterized by being a border department with an area of 20,848 km2 and to the north it has 354 kilometers of coastline [11].

La Guajira is the fifth largest department in the Caribbean region, representing 15.7% of the region's area (132,083 km2) and 1.8% of the country's area (1,140,667 km2) [12]. It is between 10°23' and 12°28' north latitude and 71°06' and 73°39' west longitude, with temperatures ranging between 22°C and 40°C [12]. It is divided into 15 municipalities and 142 populated centers between police supervision, townships, and villages, organized into three sub-regions known as Alta, Media and Baja Guajira.

Some towns belonging to Alta Guajira currently suffer from problems due to lack of electricity. Then renewable energies are a possible solution to the energy problem. Many of the rural areas are far from the urban area with complicated access roads. For this reason, there is no connection to the network and these areas do not have electricity service. Some example communities are Nazareth and Puerto Estrella, which are located in the municipality of Uribia in La Guajira, more than eight hours away by land.

Over the years the population of La Guajira has been increasing considerably, studies carried out by DANE record important figures in population growth. For the year 2008, the records estimate 763,439 inhabitants, of which 351,182 were indigenous. For 2016 there is evidence of a considerable increase of 985,452 people, which is equivalent to 1.99% of thc national population, with indigenous people occupying 46% of the population, Afro-descendants 8.2% and without ethnicity 45.9 % [13].

The Institute for Planning and Promotion of Energy Solutions for Non-Interconnected Areas (IPSE), is the organization that meets energy needs by identifying, implementing and monitoring sustainable energy solutions, with criteria of efficiency and effectiveness in Non-Interconnected Areas - ZNI, improving the living conditions of its inhabitants. The IPSE is a Public Institution of National Order, which provides its services in Colombia, with its own assets and administrative autonomy.

2. Methodology

The main objective of the study is the optimization of the hybrid system installed in the community of Nazareth, to solve the energy problem that occurs in the towns of Nazareth and Puerto Estrella, using the HOMER Legacy software as a design tool.

The hybrid system supplies energy 5 hours a day to the community, because the generator set (Diesel generator) is not in operation due to lack of fuel, which indicates that the photovoltaic array is responsible for supplying energy, the problem It lies in the lack of resources for the purchase and transfer of fuel in the system's facilities. The study focuses on the community of Nazareth. Located a few meters from the Serranía de la Macuira with a Latitude of 12.1833 and a Longitude of -71.2833 with a height of 77 meters above sea level. It is a territorial settlement 8 hours from Uribia by land with a 204 km journey. Figure 1 shows the geographical location of the town of Nazareth.

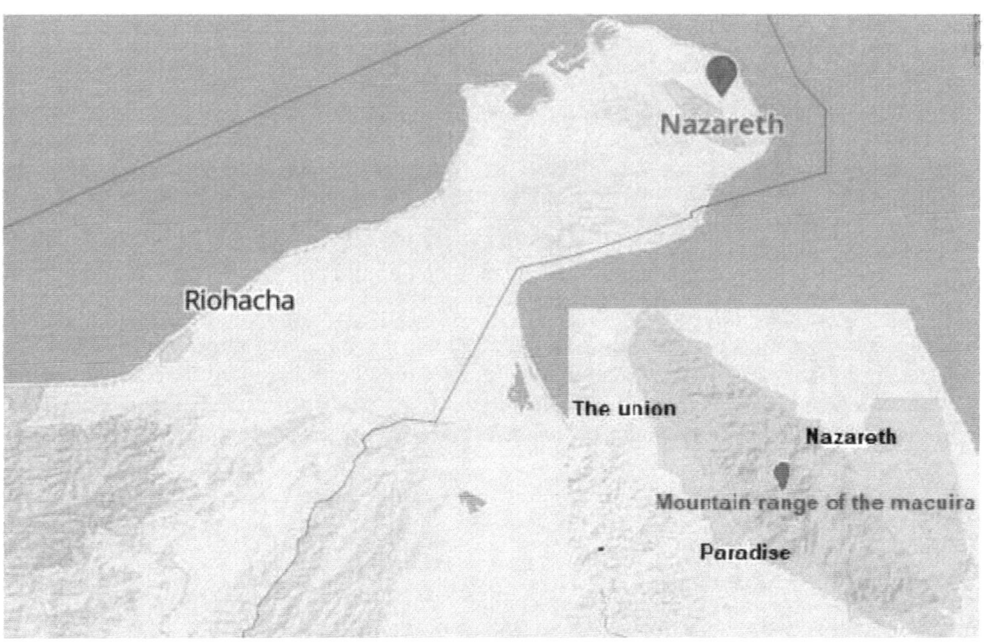

Figure 1. Location of the site: Nazareth, Guajira. (Taken from Google Map)

IOP Conf. Series: Materials Science and Engineering 1299 (2024) 012004 doi:10.1088/1757-899X/1299/1/012004

The production data and the technical characteristics of the system were supplied directly by the Institute for Planning and Promotion of Solutions for Non-Interconnected Areas (IPSE), information found in its databases, subsequently, the climatological and geographical analysis of the location, information that was provided by the National Monitoring Center (CNM).

2.1. Population demand

The register of the energy demanded by the population is determined by knowing the average number of people in the place and the possible energy consumed by each dwelling. The study carried out by the National Administrative Department of Statistics (DANE) states that between the towns of Nazareth and Puerto Estrella there is an average of 2000 inhabitants [13]. Based on the reports provided by the IPSE, the system installed in the community of Nazareth provides service to more than 250 users in Nazareth and 200 in Puerto Estrella, in addition, it reports the daily average of the energy demanded by the community, which is of 3,931 kWh/day as shown in Figure 2.

Therefore, to characterize the hourly demand profile, the average number of hours per day in which the installed system supplies electricity is defined. To determine the hours of operation of the system, the data compiled by the IPSE is used, where it establishes that the system supplies energy to Nazareth and Puerto Estrella approximately 5 hours a day. Once the average number of daily hours is obtained, the consumed energy profile is defined.

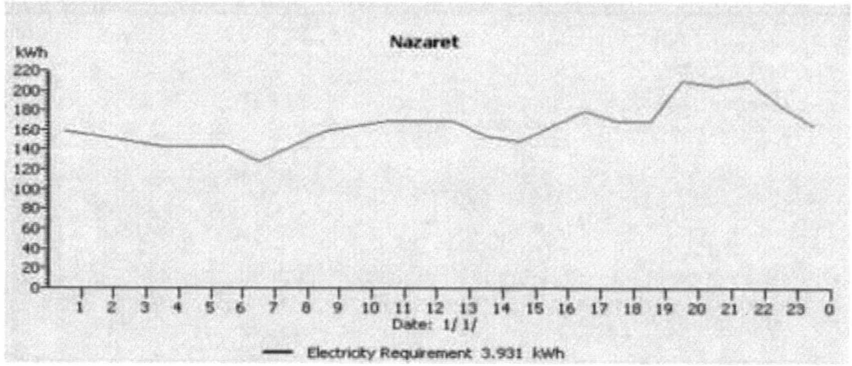

Figure 2. Daily demand profile. (Taken, with permission, from IPSE, 2017)

Since the operating time of the system only goes back to an average of five hours per day, there is no total data on the energy demanded by the community. For this reason, we proceed to interpolate the actual production data of the system with the hours of operation. To determine the total demand of the population for all months. Table 1 shows the results obtained in the study, where the monthly and annual average demand is shown.

Table 1. Monthly demand of the community of Nazareth and Puerto Estrella

Months	Daily demand (24h) kWh/day	Monthly demand kWh	Months	Daily demand (24h) kWh/day	Monthly demand kWh
January	3.168	95.040	July	3.352	100.559
February	3.429	102.891	August	3.382	101.486
March	3.328	99.863	September	3.508	105.243
April	3.366	100.988	October	3.330	99.913
May	3.262	97.883	November	3.018	90.545
June	3.223	96.706	December	2.997	89.927

Figure 3 shows the population demand profile, where three energy peaks are evidenced where more energyis demanded, being between 200kWh and 260 kWh.

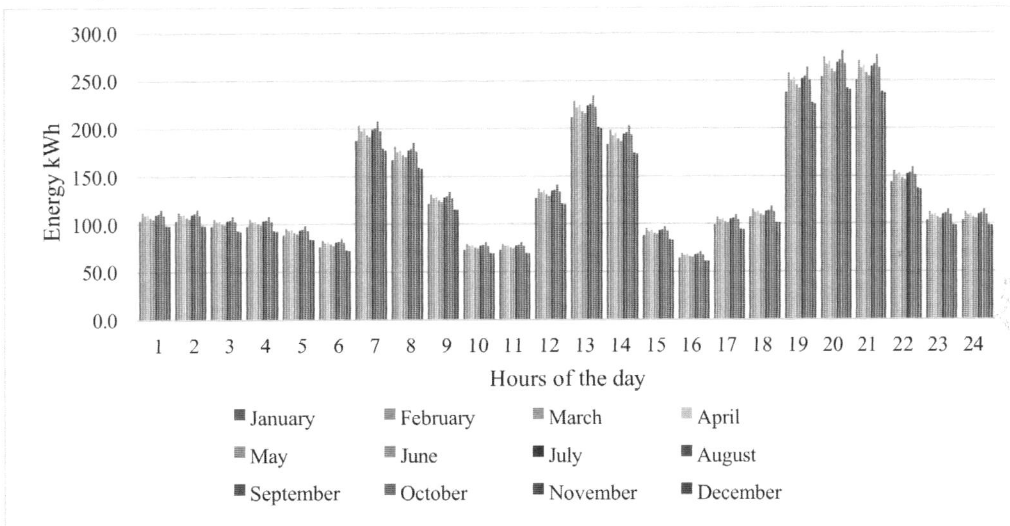

Figure 3. Hourly demand profile of the communities of Nazareth and Puerto Estrella

2.2. System simulation in HOMER software

For the simulation in HOMER, the following parameters were defined: the population demand profile (kW), monthly average wind speeds (m/s) and solar radiation (kwh/m2/day) for a year, altitude of the place (masl), solar panel capacity (kW), generator set capacity (kW), wind turbine power (kW), converter power (kW), storage system sizing.

Internally HOMER found the combination of components that satisfy the electrical loads, simulated thousands of system configurations, and generated analysis results for the input data. He made an energy balance for each of the 8,760 hours of the year. Comparing the electrical load with the energy that the system can deliver in one hour. In addition, it analyzed how each component of the system operates and the state of loading and unloading of the storage system. Figure 4 shows the main HOMER Energy window.

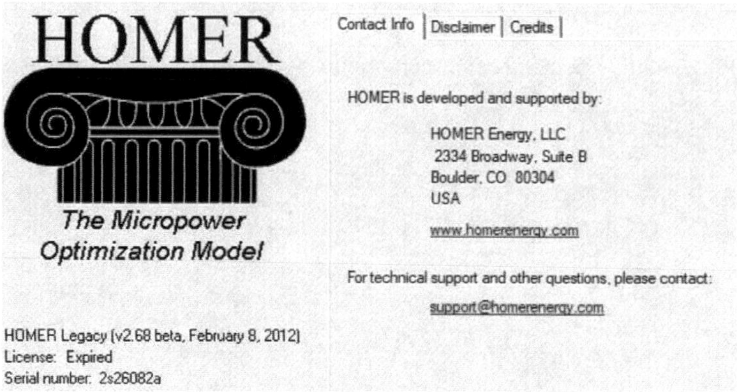

Figure 4. HOMER main window

2.2.1. Case 1: System Description

The National Monitoring Center for research provided data on wind and solar potential for the years 2017 and 2018 of the community of Nazareth, the IPSE provided reports on the energy production of the system, the information showed the daily average of energy demanded, the which was compared with the production data and the established population demand profile. The system is made up of the 320 kWp photovoltaic array, the storage system of 480 batteries each of 3350 Ah 2V, 30 Sunny Island 8.0H inverters, 28 Sunny tripower 12000 TL 10 inverters and a 364kW diesel generator set. nominal. In addition, it has a three-phase interconnection line between Nazareth and Puerto Estrella of 18.5 km and a voltage level of 13.2 kV.

The storage system is made up of the charging inverters and the battery bank, in total there are 30 inverters divided into 10 branches. Each branch is made up of three investors. Each group of inverters is connected to a circuit of two parallel battery branches, with 24 cells each (48 cells in total). Which is equivalent to a total of 20 battery strings. It should be noted that in scenario B the same system is proposed without the operation of the generator set.

Figure 5 shows the electrical plan of the system proposed in scenario A, showing the dimensioning of the main components and the power distribution, achieving a complete visualization of the system in the simplest way.

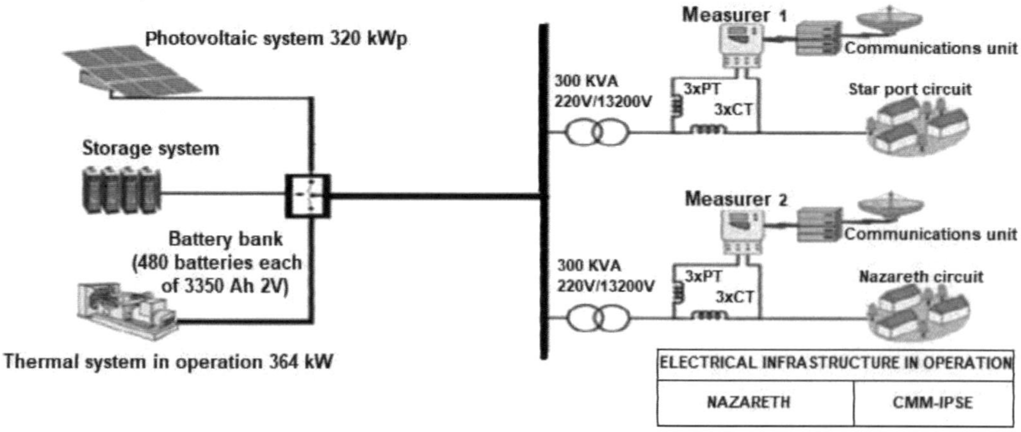

Figure 5. Diagram of the system installed in Nazareth

2.2.2. *Case 2: System Description, Scenario C*

Knowing the configuration of the system installed in Nazareth, a search space is defined in HOMER, expanding the photovoltaic matrix to 1,735 panels, 912 batteries and three wind turbines of 100 kW each, maintaining the same Diesel generator. Table 2 shows all the components of the system, the capacity, the brand and the number of equipment.

Table 2. Description of the components of scenario C

Equipment			Number of teams
Brand	**Engine**	**Ability**	
SDMO (Generador)		364 Kw	1
Solar system		434 kWp	1.735 paneles
Wind turbines		100kW	3
Battery bank (912 baterías cada una de 3350 Ah 2V)			
Sunny island 8.0H + multicluster box		6 kW	97
Sunny tripower 12000 TL10		6 kW	95
MAGNETRON (Transformador)		300 kVA	
RYMEL (Transformador)		300 kVA	

Figure 6 shows the electrical plan of the simulated system in HOMER Energy referring to scenario C, where the sizing and parts of the main components are shown, the distribution of power from the source to the utilization equipment, in order to achieve a complete visualization of the system in the simplest way.

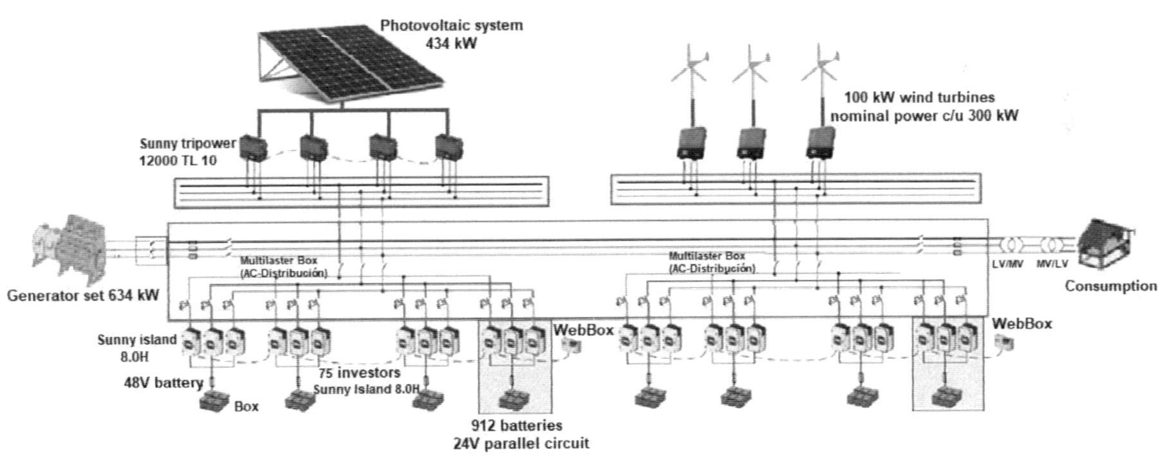

Figure 6. Single-line diagram of the simulated system at HOMER Energy, scenario C

2.2.3. *Case 2: System Description, Scenario D*

Due to the desire to reduce the use of Diesel in the system, the search space is expanded, leaving the system composed of the photovoltaic matrix system of 2,017 panels, 1,248 batteries and four wind turbines of 100 kW each. Table 3 shows the configuration, capacity and number of equipment proposed in scenario D.

Table 3. observe the configuration, capacity and number of equipment proposed in scenario D

Equipment			Number of teams
Brand	Engine	Ability	
Solar system		571 kWp	2.017 panels
Wind turbines		100kW	4
Battery bank (1.248 baterías cada una de 3350 Ah 2V)			
Sunny island 8.0H + multicluster box		6 kW	99
Sunny tripower 12000 TL10		6 kW	98
MAGNETRON (Transformador)		300 kVA	
RYMEL (Transformador)		300 kVA	

3. Results

Through the simulation carried out in HOMER, the energy production of the system referring to case 1 is determined. In scenario A, the system has a total production of 1,636,036 kWh/year, 28% of the production is from the photovoltaic array and 72% for the generator set, supplying the total demand of the population, the excess energy evidenced is 0.39%. The generator set consumes 401,177 L/year of Diesel fuel, which is equivalent to 3,607 h/year of operation. Table 4 compares the monthly production data for scenarios A and B, broken down by component.

Table 4. Comparison of production data for scenarios A and B

Energy production (scenario A)				Energy production (scenario B)
Months	FV (kW)	Generator (kW)	Total	FV (kW)
January	37.112	97.379	134.492	37.112
February	34.127	91.100	125.227	34.127
March	40.611	103.543	144.155	40.611
April	36.980	101.610	138.590	36.980
May	34.790	99.417	134.207	34.790
June	36.808	97.066	133.874	36.808
July	38.609	101.606	140.215	38.609
August	41.889	107.261	149.149	41.889
September	38.221	105.476	143.698	38.221
October	35.542	104.420	139.962	35.542
November	39.403	84.492	123.896	39.403
December	37.403	91.169	128.572	37.403
Total	451.496	1.184.540	1.636.036	451.496

3.1. Optimal system, case 2

In scenario C, the total production of the system is 1,921,545 kWh/year, 32% of the total production corresponds to the photovoltaic array, 58% to the wind system and the remaining 10% to the generator set, with an amount of fuel consumed of 64,166 L/year and 603 h/year of operation, showing an excess of energy of 13%. Figure 7 shows the monthly average of energy produced by each technology discriminated by colors.

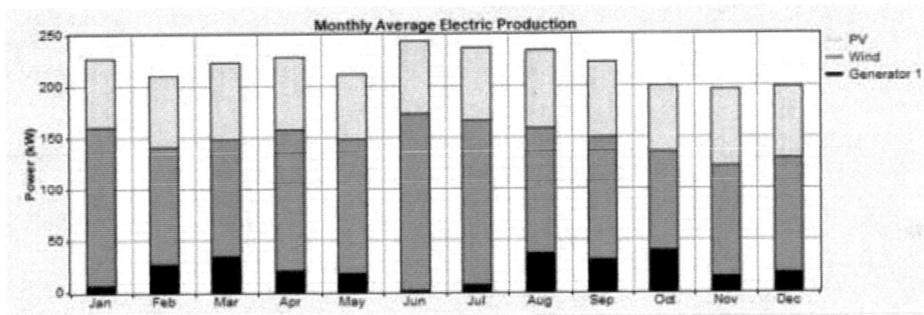

Figure 7. Monthly average of energy produced by the system

As seen in figure 8, the PV and wind systems supply the demand most of the time. The generator set runs for a few hours per day, indicating that the system's reliance on diesel power has been significantly reduced. The state of charge of the system shows 593,078 KWh/year of input energy and 511,063 KWh/year of output energy, with a range of 23.4 h and a nominal capacity used of 3,830 KWh. Figure 8 shows the hourly profile of energy produced by each system component for two days in January.

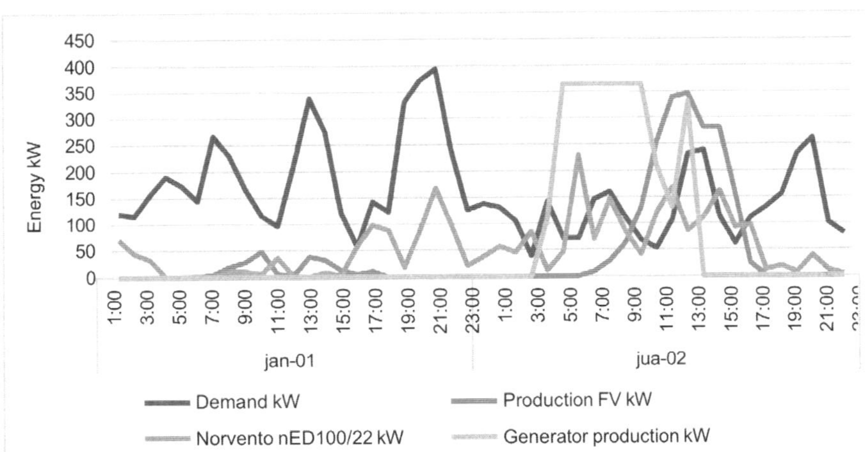

Figure 8. Hourly profile of energy produced by each component of the system for two days

In scenario D, energy production of 2,302,734 kWh/year, the wind system contributes 65% of the total production and the photovoltaic array 35%. Presenting an excess of energy of 28.4%, which represents 877,892kWh/year. Figure 9 shows the energy profile produced by components.

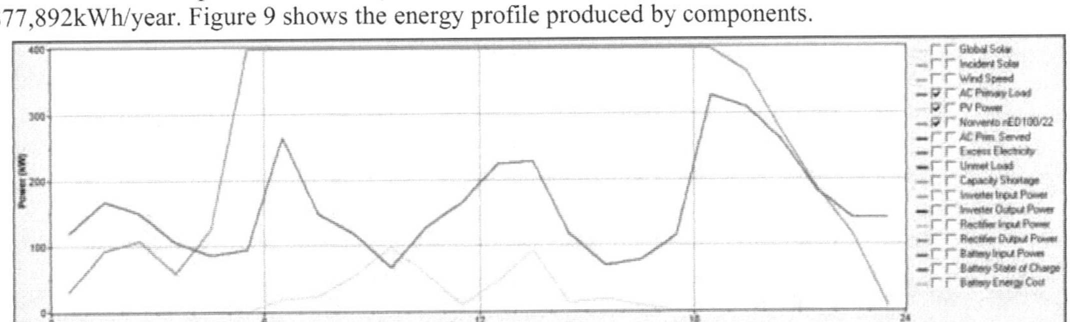

Figure 9. Profile of energy produced by components

The storage system is robust, which has an autonomy of 32 hours, with a nominal capacity of 7,408 kWh, usable nominal capacity of 5,242 kWh. Figure 10 shows the state of charge of the storage system throughout the year.

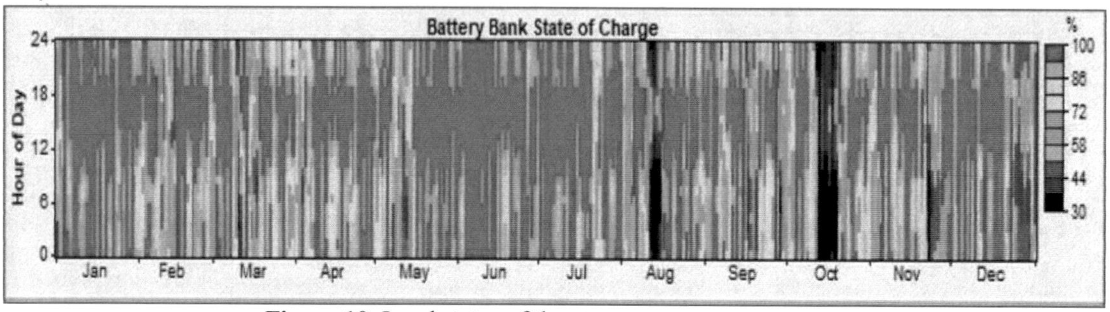

Figure 10. Load status of the storage system.

4. Discussion

This study proposes a solution to the energy problems suffered by the towns of Nazareth and Puerto Estrella, which focuses on the lack of resources of the municipal entity (Mayor's Office of Uribia) for the monthly purchase and transportation of Diesel fuel required by the plant for its operation. The results obtained from the simulated system in scenario A, when compared with the energy production of the system provided by the IPSE (scenario B), show a considerable energy deficit and the importance of the generator set for the optimal functioning of the system.

However, when the system is optimized in scenario C, it shows a favorable scenario to solve the problem, considerably reducing the energy production of the generator set. In scenario D, an alternative solution is proposed where the generator set only supports the system if a fault occurs. Comparing the data obtained in scenarios C and D, energy production satisfies the population's demand.

5. Conclusions

The simulations carried out made it possible to identify and analyze the energy problems of the communities of Nazareth and Puerto Estrella. Through the processing of wind and solar data from the location, the energy potential was satisfactorily determined, averaging 8,760 data and identifying the months and hours of the year 2017 where there is the greatest resource. It was useful to determine the population demand profile for indigenous dwellings, identifying the hours when there is the greatest energy demand.

With the optimization carried out in scenario C, the main advantage it has with respect to the system in scenario A. It is the reduction of the dependence of the system on generating energy through diesel fuel, due to the integration of wind technology, the expansion of the photovoltaic array and the battery bank.

On the other hand, the configuration proposed in scenario D is a tempting alternative, because it would not be necessary to have monthly resources for the purchase and transportation of fuel, another aspect to consider is the dimensioning of the system, which is more robust compared to the one proposed. in scenario C, the surplus of unusable energy is added to this. Therefore, the optimal configuration is the one proposed in scenario C, where the generator set only operates 603 h/year, considerably reducing fuel consumption, which leads to having few resources for its acquisition, the system has a 90% capacity. autonomy without the generator set, carrying out a controlled supply plan, the energy conditions of the towns of Nazareth and Puerto Estrella would change positively.

Acknowledgments: I thank the University of La Guajira for having the necessary resources to finance the research, the Institute for Planning and Promotion of Energy Solutions for Non-Interconnected Areas (IPSE) and the National Monitoring Center (CNM) for all the provided information provided in its databases. To the Center for Energy, Environmental and Technological Research (CIEMAT) and to the Carlos III University of Spain for accompaniment.

References

[1] X. Xiaofeng, W. Zhifei, J. Qiang, W. Chenglong y G. Guowei, «Global renewable energy development: Influencing factors, trend predictions and countermeasures,» ELSEVIER , 2019.

[2] C. y. A. E. Gasca, «Transición energética, energías renovables y energía solar de potencia,» Revista Mexicana de Física, pp. 75-84, 2013.

[3] C. R. Algarin y O. R. Álvarez, «Un panorama de las energias renovables en el Mundo, Latinoamerica y Colombia,» 2018. [En línea]. Available:https://www.revistaespacios.com/a18v39n34/a18v39n34p10.pdf.

[4] F. Pazheri y M. H. Othman, «A review on global renewable electricity scenario,» *ELSERVIER,* p. 31, 2014.

[5] H. G. A. Corredor, «Análisis costo beneficio de energías renovables no convencionales en Colombia,» *Fedesarrollo, Centro de Investigación Económica y Social ,* pp. 1-90, 2013.

[6] J. Caspary, «Gauging the future competitiveness of renewable energy in Colombia,» *ELSEVIER ,* p. 31, 2009.

[7] M. d. E. UPME., «Formulación de un plan de desarrollo para las fuentes no convencionales de energiaen Colombia (PDFNCE).,Plan Energético Nacional,» CorpoEma.,, Colombia, 2010.,2019.

[8] H. Benavides y O. Sinbaqueva, «SIMEC,» 18 Marzo 2022. [En línea]. Available:http://www.si3ea.gov.co/Home/EnergiaSolar/tabid/74/language/en-US/Default.aspx.

[9] A. Pinilla y L. Rodriguez, «Evaluación de desempeño del Parque Eólico Jepirachi,» *EL SEVIER ,* pp.48-52, 2009.

[10] CIEMAT , «Sistemas eólicos distribuidos y aislados,» Madrid, España, CIEMAT, pp. 1-68.

[11] M. Vanegas y E. villicaña, «Cuantificación y caracterización de la radiación solar en el departamento de La Guajira-Colombia mediante el cálculo de transmisibilidad atmosférica,» *Prospectiva ,* pp. 1-10,2015.

[12] L. Puerta y O. Cetina, «Problemas y Soluciones Ambientales,» 5 Noviembre 2013. [En línea]. Available: http://problematicasysolucionesambientales.blogspot.com/2013/11/generacion-de- energia-en-el.html..

[13] A. Martinez, P. Medina y J. Pachón, «La Guajira, Caracterización Departamental y Municipal,» Fedesarrollo, La Guajira, 2019.

[14] C. Y. R. O. Robles, «Un panorama de las energías renovables en el mundo, Latinoamerica y Colombia,» *Espacios,* p. 16, 2018.

Characteristics in the Fourier spectrum of images associated with discontinuities such as cracks and fissures in concrete structures.

C L Sandoval-Rodriguez [1], B E Tarazona-Romero[1], O Lengerke[1], A D RinconQuintero[1], K T Jaimes-Quintero[1].

[1] Unidades Tecnológicas de Santander UTS, Bucaramanga Santander 680005, Colombia

Csandoval@correo.uts.edu.co, Btarazona@correo.uts.edu.co, Olenguerke@correo.uts.edu.co, Aricon@correo.uts.edu.co and Ktjaimes@uts.edu.co

Abstract. Increasing An algorithm is developed to identify differences between concrete slabs with some type of Crack and Fissure discontinuity. The Fourier transform is used as a basis. Different types of fil-ters were evaluated within the image processing algorithm, in order to adapt favorably to the discontinuity detection process, allowing the image data to be read in the frequency domain, generating a figure of merit to compare the image. where there are no discontinuities and those that have some alteration in their structure.

Keywords: Image processing, Fourier transform, Non-destructive testing, Concrete, Cracks and fissures.

1. Introduction

Template Concrete is the most frequently used building material to date, contributing to the development of infrastructure worldwide. However, this material presents damage due to being subjected to stresses upper its resistance capacity, generating anomalies known as "cracks" and "fissures" type discontinuities. These phenomena severely affect the appearance, life, and intrinsic properties of concrete. [1]. Various causes generate cracks and fissures, although they can be grouped into two types of actions: mechanical and spontaneous [2].

In addition, these discontinuities can be generated during the concrete setting or hardening stages, generating a greater or lesser effect on the hardness and resistance of the material [3]. Triggering a variation in the material and causing the cracks and fissures to have a level of greater severity for the integrity of the concrete; evident in the dimensions of the discontinuity, especially in the width [4]. Therefore, it is interesting to identify early, both discontinuity and causes, to keep impermeability and limit the damage. [5] [6].

Currently, crack detection is mostly done in two ways: applying destructive and non-destructive testing. [7]. Non-destructive testing excels, due to its ease of application and its reliability in results. The use of visual tools, application of laser equipment, ultrasound [8], ray, infrared and thermal signals [9], They are part of some non-destructive tests that are used more frequently at the laboratory and industrial level [10].

Additionally, the methodology for detecting crack-type discontinuities and fissures is currently being applied through image processing. [11]. Although this technique is rarely used, due to difficulties generated by various reasons such as the random shape and irregular size of the cracks, irregular lighting conditions, shadows, imperfections, and flaking in the concrete of the acquired images, it is projected as an alternative. for the detection of these types of anomalies [12] [13].

Consequently, the methods based on image processing are summarized and divided into four categories: integrated algorithm, morphological approach, method based on percolation and practical technique. [14] [15]. Although the use of image processing is rare, there are advances that seek to develop an artificial vision system that serves as a tool capable of enhancing images, previously taken, and create a new one in which the discontinuities present in the image are more clearly evident. initial image [16] [17] As in turn, other systems aim to improve certain aspects of the images, to achieve highlighting specific details [18]. Thus, this document presents the development of an algorithm for the detection of discontinuities such as cracks and fissures, through the image processing technique applying the Fourier transform as an alternative for identification based on non-destructive tests of anomalies in structures of concrete.

2. Materials and methods

This work is develops using the analytical method, where the information of a series of images is observed and examined using the Fourier transform, as a method of converting the image information to the frequency domain; using the MATLAB system as a programming tool, which allows obtaining data sets in matrices where the information is stored and in turn, using this information for the development of the algorithm, for the comparison of possible alterations at the end of the method [19]. By enhancing the con-version of the information, a comparative analysis was carried out between the results of an image where discontinuities were evident and one where the plate was free of these.

The information from the sample images is stored and compared with an original basis to determine differences and observe the effect it can have on other variables. Results are grouped and represented in a numerical way that shows the quantitative approach.

The purpose of the algorithm is to determine characteristics in the frequency domain. These features can be associated with discontinuities. We seek to obtain information that describes the phenomenon and not why it happens is. Figure 1 shows the process used in the algorithm that allows obtaining information generated by the discontinuities in the images.

For the development of the algorithm, we rely on the Fourier transform, known the frequency spectrum of a function. An example of this is what the human ear does since it receives waves and allows decomposition in different frequencies (that can to heard). The human ear perceives different frequencies as time passes. However, the Fourier transform contains all the frequencies contained in all the times in which the signal existed; that is, in the Fourier transform a single frequency spectrum obtained for the entire function [20]. The Fourier transform is an application that maps a function (f), with complex values and defined on the line, with another function (g) defined as follows (equation 1):

$$F(\omega) = \frac{1}{2\pi}\int_{-\infty}^{\infty} f(x)e^{i\omega x}dx \tag{1}$$

MATLAB was needed to be able to choose commands compatible with the Fourier transform and that in turn allows us to facilitate the reading of images and their treatment. The intention of this image processing is to increase the signal-to-noise ratio and accentuate the characteristics of the images using custom filters or predefined by MATLAB. Now, MATLAB defines the image as a two-dimensional function f (x) where (x) and (y) are the spatial coordinates, and the value of (f) in any pair of coordinates (x) is the intensity of the image at that point. An image can be continuous with respect to (x) and (y), and also in intensity (analog image). Converting this image to digital format requires both the coordinates and the intensity to be digitized. Digitizing the coordinates is called sampling while digitizing the intensity is called quantization. So, when all quantities are discrete, we call the image a digital image. [11]

Initially, an analysis using Fast Fourier transformation (FFT) was performed. The objective is to compare these results using a series of commands that allow the information to storage in matrices made up of real numbers [21]. This makes it easy to select the filter that best suits the needs of the algorithm; since this technique allows to improvement or modify an image. For example, you can filter images to emphasize certain entities or remove other entities. Image processing operations implemented with filtering include anti-aliasing, sharpening, and edge enhancement. [22].

The filter used in the algorithm is a two-dimensional filtering method that uses the inverse Fourier transformation to generate a response at the desired frequency. It is a matrix containing the desired frequency response at equally spaced points on the Cartesian plane. The equation 2 describes the filtering. [23]:

$$h_d(n_1, n_2) = \frac{1}{(2\pi)^2}\int_{-\pi}^{\pi}\int_{-\pi}^{\pi} H_d(\omega_1, \omega_2) e^{j\omega_1 n_1} e^{j\omega_2 n_2} d\omega_1 d\omega_2$$

$$h_d(n_1, n_2) = h_d(n_1, n_2)\omega(n_1, n_2) \quad (2)$$

Where: Hd is the filter frequency response.

Figure 1. Image discontinuity detection process.

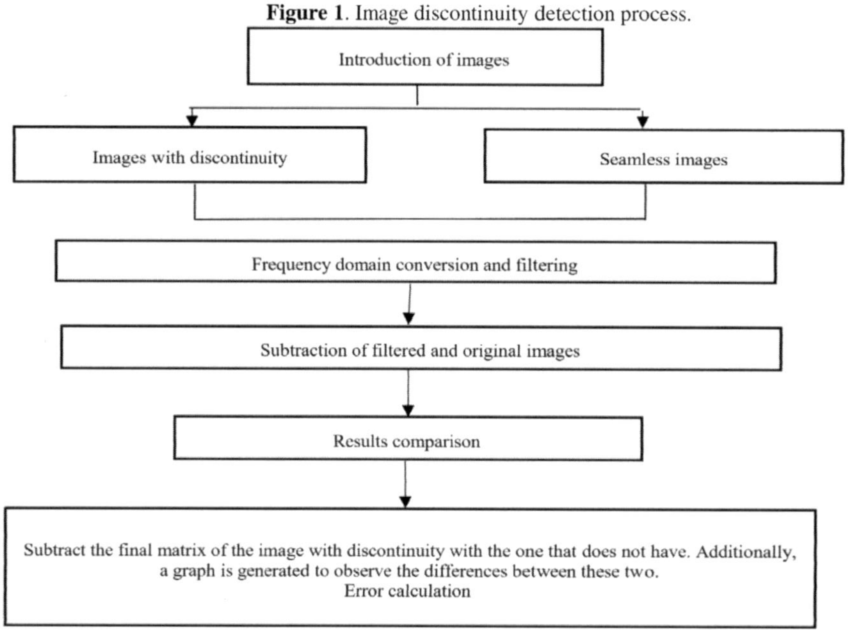

Source: Author

Finally, as a method of comparison of the elements obtained when filtering the images, the use of the mean square error (MSE) was used, this is defined as the way to evaluate the difference between an estimator and the real value of the quantity that you want to calculate. The MSE measures the average of the square of the "error", the error being the value by which the estimator differs from the amount to be estimated. Where, Ŷ is a predictor vector with n elements and Y is the really values vector [24] (see equation 3).

$$MSE = \frac{1}{n}\sum_{i=1}^{n}(\hat{Y}_i - Y_i)^2 \quad (3)$$

Finally, as a method of comparison of the elements obtained when filtering the images, the use of the mean square error (MSE) was used, this is defined as the way to evaluate the difference between an estimator and the real value of the quantity that you want to calculate. The MSE measures the average of the square of the "error", the error being the value by which the estimator differs from the amount to be estimated. Where, $\hat{Y}$ is a predictor vector with n elements and Y is the really values vector [24] (see equation 3).

The information obtained through the Fourier transform in conjunction with the MSE improves the results obtained by filtering and thus achieves the best possible pattern classification [25].

3. Results

The algorithm employs the FFT to detect differences in features associated with cracks and fissures in concrete structures. Different Commands, which together allowed us to filter the images and to see their information in the frequency domain. In fig 2 it can see three examples using the propose algorithm.

Figure 2. (a) Filtering result plate 1 (frequency domain). (b). Filtering result plate 2 (frequency domain). (c) Filtering result plate without discontinuity (frequency domain)

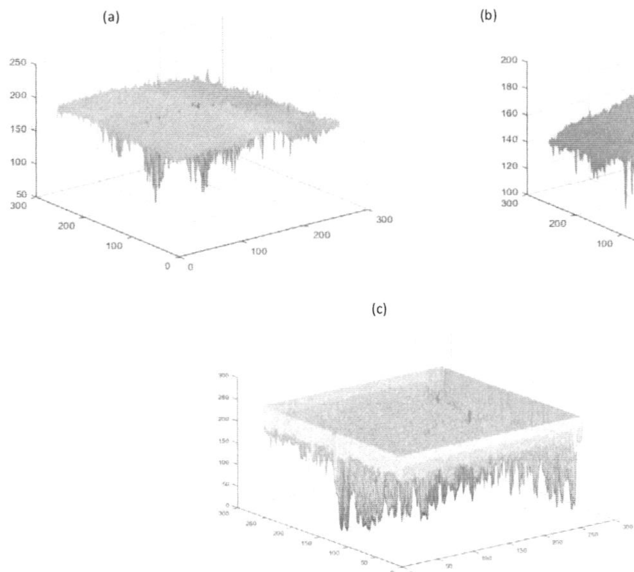

Source: Author

In Figs 4-5, examples of the studied plates, and plate of reference are shown. With this information, the existence of discontinuities in them is evidence. However, the amount and diversity of information obtained in each resulting image did not identify the characteristics of the cracks or fissures.

Figure 3. Results (plate without discontinuity vs others plates (calculations in the frequency domain).

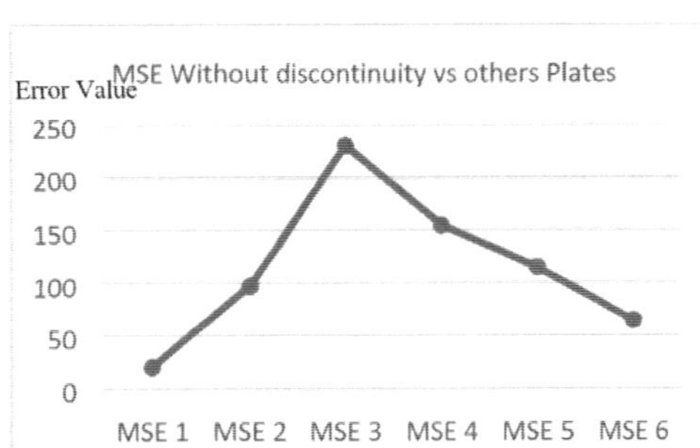

Source: Author

Figure 3 shows the error value obtained when comparing the data obtained from the image that shows discontinuities and one in which it does not (which is the same for all comparisons). Each point refers to comparison, and the error value is interpreted as the deviation of the data with respect to the discontinuous image from the non-discontinuous one.

Thanks to the results obtained by the algorithm, it was possible to determine the dis-continuities in the images. However, the information is not enough to determine a specific range to classify cracks or fissures. If we can group them by the resulting error (figure 3). It can be seen that plates 3 and 4 are the greater clarity, therefore with the greatest error. This is due to the fact that the algorithm compares each plate with the one that has no discontinuity and has a darker contrast. On the other hand, there are plates 1 and 6 which are similar in contrast and brightness to the plate of figure 4. (e), for which a smaller error is obtained.

Figure 4. (a) Plate 1. (b) Plate 2-crack. (c) Plate 3. (d) Plate 4. (e) Plate 5. (f) Plate 6.

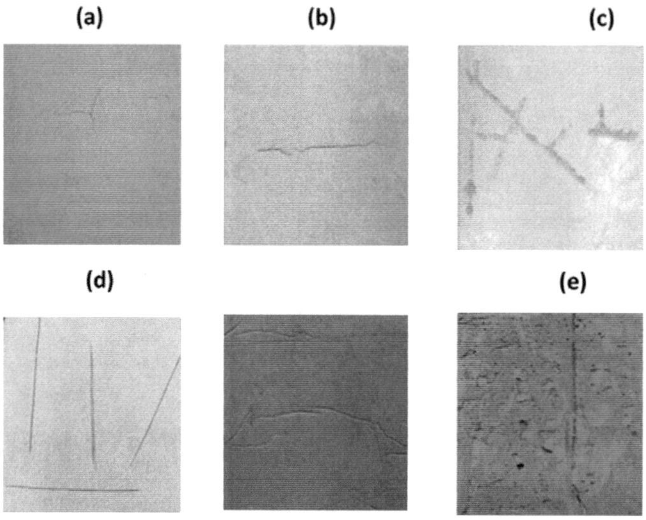

Source: Author

Figure 5. Plate without discontinuity.

According to raise above, the quality and clarity of information are important features. A factor like the illumination, location of the capture, and the brightness generates variation in the error. The followings figures contain all information necessary for the study.

4. Discussion

The proposed hypothesis focuses on deciding whether the Fourier spectrum of the image can provide information that allows differentiating the types of structural discontinuities studied. Differences are observed that have been quantified by FFT, frequency filtering, and MSE calculation. With this methodology, it is possible to classify between structures without alterations and with alterations. However, the quantification of the damage (differentiating between crack or fissure) is not achieved with the present work.

Other works have implemented artificial intelligence tools, where the prediction capacity is evaluated but from the perspective of the classifier and not from its characteristics [4], [26] and [16]. In our work, we present the calculation of one feature like MSE how a descriptor of the structural condition. However, it was observed that other factors affect the MSE such as lighting, location, and brightness that are associated with the way the image is captured. In previous works [17], [18] already exists some treatment employing other techniques like mechanic vibrations-wavelet, and imaging processing with special techniques that shown good results, but nor does it quantify the magnitude of the damage; in our work it is possible quantify the differences between both patterns (cracks and fissures). Although these same factors (lighting, location, and brightness) generate a data variation that does not allow to associate specific characteristics with cracks or fissures, when performing the process several times with different images, the result indicates that in images that have a similar contrast, the error can be classified into a determined range.

The same behavior occurs with better-quality images. In general, if the images that contain the discontinuities have a contrast similar to that of the base image, the error will be minimal, on the other hand, if it is of light tonality, contrary to the base image, the margin of error is high. These conditions are what make it possible to demonstrate the need to standardize the images and thus create the least alteration of the information.

5. Conclusions

An algorithm capable of extracting relevant information to quantify the differences between a plate with some alterations has been proposed. The standardization of test im-ages is left for future work, using digital image processing tools.

References

[1] J. Hernández, «The approach of the ACI-318 for crack control. A vision from the Cuban reality», *Revista de la construcción*, vol. 13, n.º 2, pp. 15-21, ago. 2014, doi: 10.4067/S0718915X2014000200002.

[2] J. T. Corral, «Patología de la construcción grietas y fisuras en obras de hormigón ; origen y prevención», *Ciencia y Sociedad*, vol. 29, n.º 1, pp. 72-114, 2004.

[3] F. Ye, G. Bianchi, S. Rane, S. A. Tassou, y J. Deng, «Analytical grid generation and numerical assessment of tip leakage flows in sliding vane rotary machines», *Advances in Engineering Software*, vol. 159, p. 103030, sep. 2021, doi: 10.1016/j.advengsoft.2021.103030.

[4] M. Flah, A. R. Suleiman, y M. L. Nehdi, «Classification and quantification of cracks in concrete structures using deep learning image-based techniques», *Cement and Concrete Composites*, vol. 114, p. 103781, nov. 2020, doi: 10.1016/j.cemconcomp.2020.103781.

[5] C. Liu y G. Liu, «Characterization of pore structure parameters of foam concrete by 3D reconstruction and image analysis», *Construction and Building Materials*, vol. 267, p. 120958, ene. 2021, doi: 10.1016/j.conbuildmat.2020.120958.

[6] W. Tian, X. Cheng, Q. Liu, C. Yu, F. Gao, y Y. Chi, «Meso-structure segmentation of concrete CT image based on mask and regional convolution neural network», *Materials & Design*, vol. 208, p. 109919, oct. 2021, doi: 10.1016/j.matdes.2021.109919.

[7] A. Goñi y J. Fernandez, «Ensayos no destructivos para localización de grietas en piezas estampadas», Trabajo Fin de Grado, UPNA, Pamplona, 2014. [En línea]. Disponible en: http://academica-e.unavarra.es/bitstream/handle/2454/11803/TFGGo%F1iCordobaAlicia.pdf?sequence=1

[8] M. Lott *et al.*, «Fracture processes imaging in concrete using nonlinear ultrasound», *NDT & E International*, vol. 120, p. 102432, jun. 2021, doi: 10.1016/j.ndteint.2021.102432.

[9] J. Gu y S. Unjoh, «Image processing methodology for detecting delaminations using infrared thermography in CFRP-jacketed concrete members by infrared thermography», *Composite Structures*, vol. 270, p. 114040, ago. 2021, doi: 10.1016/j.compstruct.2021.114040.

[10] F. E. Silva, «Detección automática de grietas basada en imágenes de puentes de concreto a través de modelos Encoder-Decoder», 2019, Accedido: 14 de julio de 2022. [En línea]. Disponible en: https://repositorio.uchile.cl/handle/2250/174437

[11] Y. Liu y J. K. W. Yeoh, «Automated crack pattern recognition from images for condition assessment of concrete structures», *Automation in Construction*, vol. 128, p. 103765, ago. 2021, doi: 10.1016/j.autcon.2021.103765.

[12] K. Dinh, N. Gucunski, K. Tran, A. Novo, y T. Nguyen, «Full-resolution 3D imaging for concrete structures with dual-polarization GPR», *Automation in Construction*, vol. 125, p. 103652, may 2021, doi: 10.1016/j.autcon.2021.103652.

[13] A. Mohan y S. Poobal, «Crack detection using image processing: A critical review and analysis», *Alexandria Engineering Journal*, vol. 57, n.º 2, pp. 787-798, jun. 2018, doi: 10.1016/j.aej.2017.01.020.

[14] M. E. Paoletti, J. M. Haut, J. Plaza, y A. Plaza, «Estudio Comparativo de Técnicas de Clasificación de Imágenes Hiperespectrales», *Revista Iberoamericana de Automática e Informática industrial*, vol. 16, n.º 2, Art. n.º 2, mar. 2019, doi: 10.4995/riai.2019.11078.

[15] M. Y. Moreno Revelo, «Estudio comparativo de técnicas de visión artificial y procesamiento de imágenes enfocadas a la detección de cambios en coberturas boscosas», *Comparative study of artificial vision techniques and image processing focused on changes detection in forest covers*, 2020, Accedido: 14 de julio de 2022. [En línea]. Disponible en: https://repositorio.unal.edu.co/handle/unal/77681

[16] W. Wei, L. Ding, H. Luo, C. Li, y G. Li, «Automated bughole detection and quality performance assessment of concrete using image processing and deep convolutional neural networks», *Construction and Building Materials*, vol. 281, p. 122576, abr. 2021, doi: 10.1016/j.conbuildmat.2021.122576.

[17] B. E. Romero-Tarazona, C. L. Rodriguez-Sandoval, J. G. Villabonai-Ascanio, y A. D. RincónQuintero, «Development of an artificial vision system that allows non-destructive testing on flat concrete slabs for surface crack detection by processing of digital images in MATLAB», *IOP Conf. Ser.: Mater. Sci. Eng.*, vol. 844, n.º 1, p. 012058, may 2020, doi: 10.1088/1757899X/844/1/012058.

[18] B. Tarazona y C. Sandoval, «Evaluation of cracks and fissures discontinuities in concrete structures using a vibration analyzer and digital image processing», *Entre Ciencia e Ingeniería*, vol. 13, n.º 25, 2019, doi: https://doi.org/10.31908/19098367.4018.

[19] B. E. Tarazona, C. L. S. R, C. G. C. Arias, J. G. A. V, y J. J. V. N, «Detection of structural alterations in metal bodies: An approximation using Fourier transform and principal component analysis (PCA)», *Scientia et Technica*, vol. 25, n.º 2, Art. n.º 2, jun. 2020, doi: 10.22517/23447214.23501.

[20] F. Vela, «Transformada de Fourier en procesamiento digital de Imágenes Funciones de Variable Compleja», p. 4, 2013.

[21] C. Sandoval Rodriguez, B. Tarazona Romero, O. Lengerke Perez, C. Cárdenas-Arias, D. C. Diaz, y O. Cárdenas, «Descriptive Study of a Rotary Machine Affected by Misalignment and Imbalance Applying the Wavelet Transform», 2021, pp. 226-242. doi: 10.1007/978-3-030-72212-8_17.

[22] «MathWorks - Creadores de MATLAB y Simulink - MATLAB y Simulink». https://es.mathworks.com/ (accedido 14 de julio de 2022).

[23] J. S. Lim, *Two-dimensional signal and image processing*. Englewood Cliffs, N.J.: Prentice Hall, 1990.

[24] R. J. Hyndman y A. B. Koehler, «Another look at measures of forecast accuracy», *International Journal of Forecasting*, vol. 22, n.º 4, pp. 679-688, oct. 2006, doi: 10.1016/j.ijforecast.2006.03.001.

[25] C. L. Sandoval Rodríguez, A. A. Barros, y S. A. Herreño Ávila, «Automatic Classification of mechanical vibration patterns in rotating machinery affected by unbalance», *INGE@UAN - TENDENCIAS EN LA INGENIERÍA; Vol. 4 Núm. 7 (2013)*, jun. 2021, Accedido: 14 de julio de 2022. [En línea]. Disponible en: http://repositorio.uan.edu.co/handle/123456789/3932

[26] M.-T. Cao, N.-M. Nguyen, K.-T. Chang, X.-L. Tran, y N.-D. Hoang, «Automatic recognition of concrete spall using image processing and metaheuristic optimized LogitBoost classification tree», *Advances in Engineering Software*, vol. 159, p. 103031, sep. 2021

Obtaining the percentages of ductility and brittleness of AISI/SAE 1020 and 304 steels, using digital image processing.

A D Rincon-Quintero[1], O Lengerke[1], C L Sandoval-Rodriguez [1], K T JaimesQuintero[1].

[1] Unidades Tecnológicas de Santander UTS, Bucaramanga Santander 680005, Colombia

Aricon@correo.uts.edu.co, Olenguerke@correo.uts.edu.co, Csandoval@correo.uts.edu.co, and Ktjaimes@uts.edu.co

Abstract. This research is the integration of a set of elements in a system of capturing, processing, and digital image analysis. It allows a better visual and numeric interpretation to determine the toughness, ductility percentage, and fragility of steel AISI/SAE 1020 and 30, getting better qualitative and quantitative observation of the results from the impact test (Charpy test). Patterns in the standard form, ASTM E23, were digitalized to evaluate the percentage of ductility/fragility of specimen testing. After, we calculated the area and the equivalent diameter of the material. using digital image processing and numerical comparison between the patterns specified in the standard form ASTM E23 and the testing in the impact test , and it allows to find the kind of pattern it is closest and determine which degree of ductility. Finally, the results were compared by three experts. The algorithm accuracy was 80%.

Keywords: Charpy Test, Digital Image processing, Case based reasoning, Pattern recognition

1. Introduction

Digital image processing has been used in multiple applications[1],[2],[3],[4]. The applications refer to topics such as fruit selection[5], classification of medical images[6],[7], structural analysis [8],[9],[10] and materials classification [11],[12]. Secondly, the Charpy test , has been used to determinate toughness[13], behavior analysis in thermic variation [14], and identification of different properties of materials[15]. For the design of machine elements, it is important to evaluate characteristics such as ductility and brittleness in order to calculate strength[16],[17]. One of the main drawbacks in determining brittleness and ductility is that it is associated with direct observation by experts through microscopic images. Human subjectivity is always a problem. In this work , we was estimated fragility and ductility using the ASTM E23 standard [18].

2. Materials and methods

2.1 Impact Charpy Machine and Specimen test

The Charpy test is used in order to determine the impact energy on a test material. This is related to the stress perceived by a test specimen, and allows to make a correct selection of material according to end use[1], [18]. Figure 1, shows the impact Charpy machine. Specimens specifications in Figure 2.

IOP Conf. Series: Materials Science and Engineering 1299 (2024) 012006 doi:10.1088/1757-899X/1299/1/012006

Figure 1. Impact Charpy Machine. Used to charpy test.

Source: Author

Figure 2. Types of notch in the Charpy test test

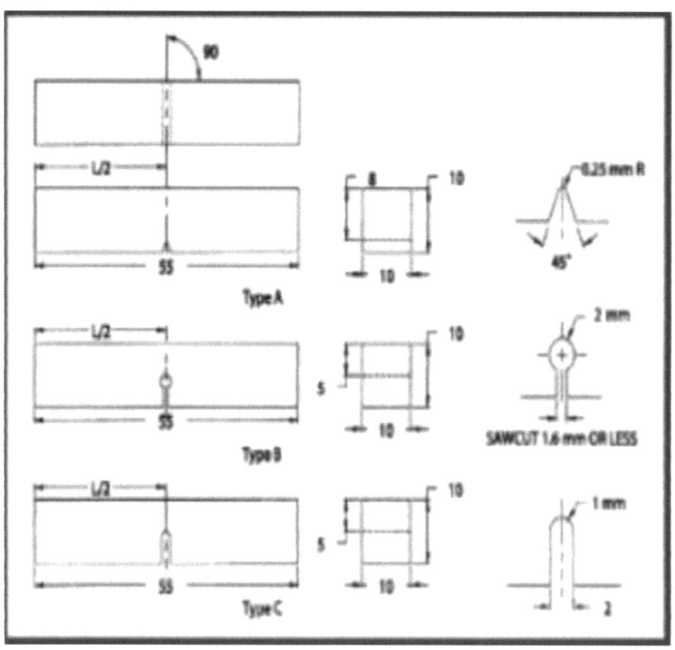

Source: Author

2.2 Methodology

AISI SAE 304 and 1020 steel specimens were used. Charpy tests were performed according to standard[18]. Afterwards, both the master images and the fractures of the specimens were digitized using a kanon camera with 8 mega pixels of resolution. To improve the definition, a preprocessing of the fractures of the specimens was necessary, which consisted of adding graphite to the surfaces of the fractures (see an example in figure 5). Finally, the application of the digital image processing algorithm (see figure 4). Figure 3 shows the flow diagram for methodology.

Figure 3. Methodology diagram

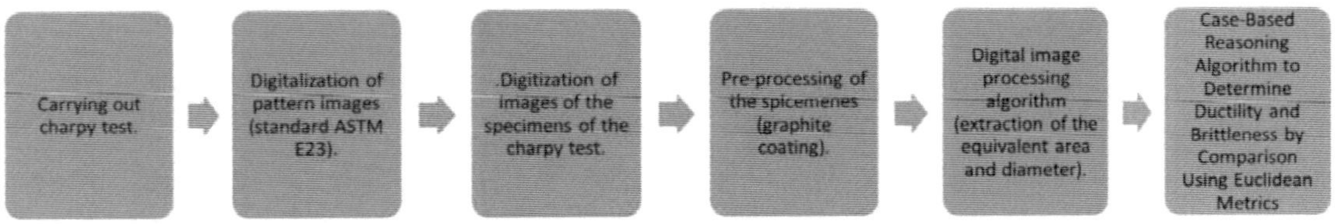

Source: Author

Figure 4. Digital image processing algorithm

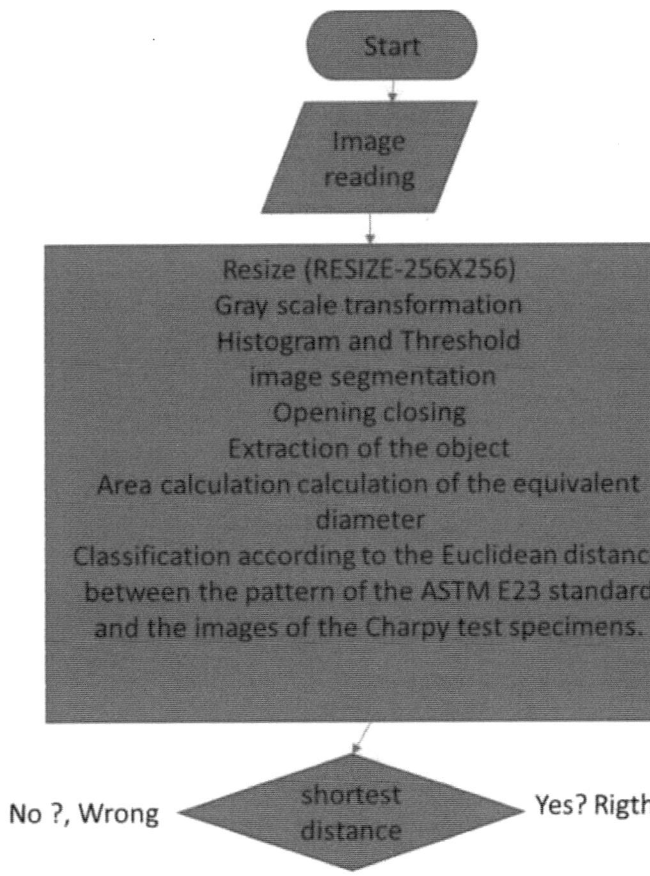

Source: Author

Figure 5. Left : AISI 304 steel wihput graphite. Right

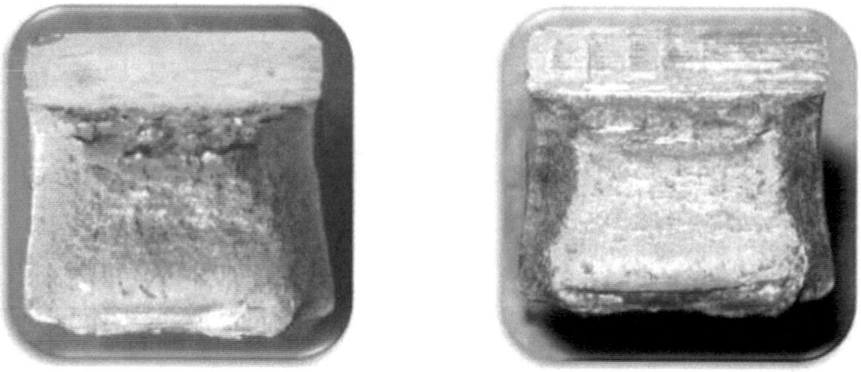

Source: Author

2.3 Digital image processing algorithm

Figure 4 shows the image processing algorithm. The process is based on morphological operations such as opening, closing and segmentation[3], [4], [19]. The objective was to obtain binarized images in order to extract characteristics such as the equivalent area and diameter[20]. Comparison was made using Euclidean distance (see equation 1) which is an tool to reasoning case based[21],[9], [22]. To obtain good results, five repetitions were necessary, according to[23]. Figure 6 and 7 shows the results of the images processing.

$$d(p1, p2) = \sqrt{(x2 - x1)^2 - (y2 - y1)^2} \quad (1)$$

Where: x2 and y2 are the pattern descriptors (equivalent area or diameter). x1 and y1 are observed values for Charpy test specimen.

3. Results

3.1 Digitization of pattern images

Figure 6 shows the patterns of the standard ASTM E23. Figure 7 shows the capture process.

Figure 6. Patterns of ASTM E23. With associated ductility percentage.

Source: Author

Figure 7. Digitization process

Source: Author

3.2 Digitization of Charpy test specimens

Figure 8 shows the digitization process of the Charpy test specimen, both AISI 304 steel (left) and AISI/SAE 1020 steel (right). Here it observes examples for 6 specimens.

Figure 8. AISI 304 steel (left) and AISI/SAE 1020 steel (right)

Source: Author

3.3 Featuring extraction (area and diameter)

In order to detect the area and equivalent diameter, we calculated this feature from binary images. Results of the process are shown in figures 9 and 10. Figure 9 upper panel (first-left) corresponds to original image in RGB format. Second-center is the histogram of Image resized (and gray-scale). In bottom panel it can see the binarized image (area) and perimeter. The equivalent diameter is obtained as: perimeter/4.

Figure 9. Processing of a pattern image using the algorithm. (Area and Equivalent diameter)

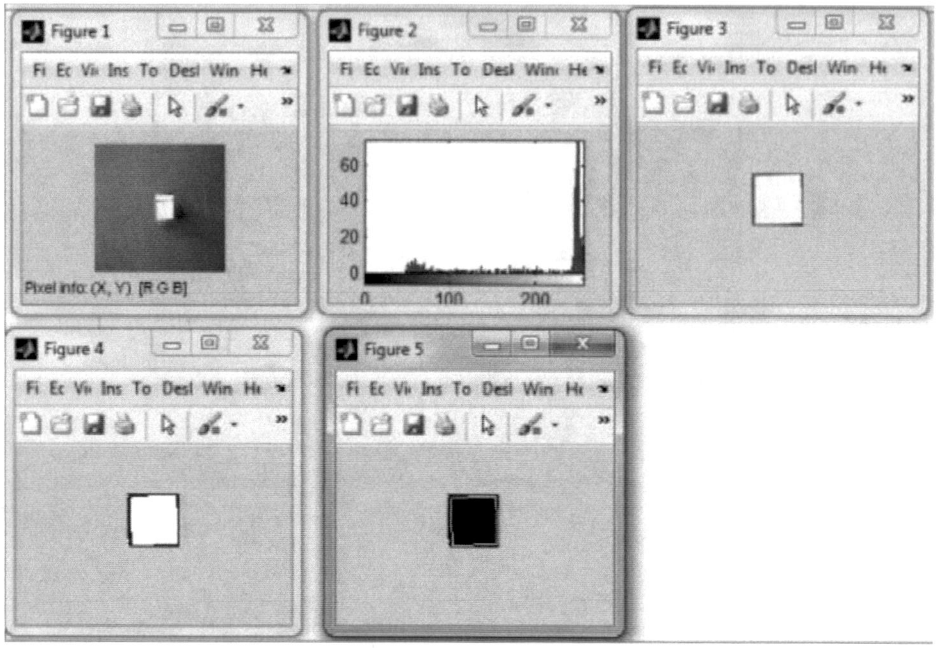

Source: Author

Figure 10 shows the obtention of equivalent diameter and area process. It can see in the bottom panel the area (center) and perimeter.

Figure 10. Processing of a Charpy test image using the algorithm. (Area and Equivalent diameter)

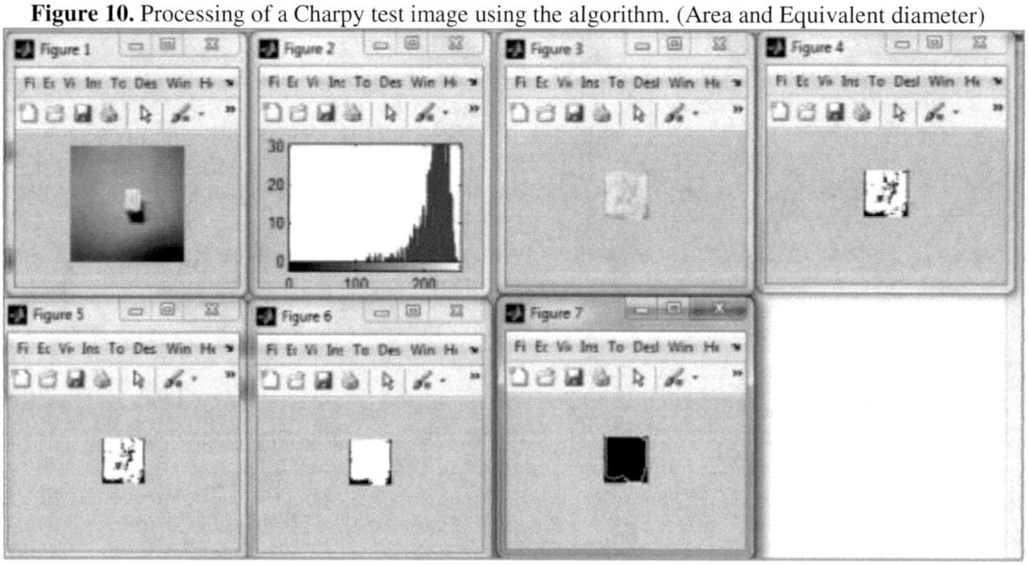

Source: Author

IOP Conf. Series: Materials Science and Engineering 1299 (2024) 012006 doi:10.1088/1757-899X/1299/1/012006

3.4 Ductility / fragility Obtention, using Reasoning case based

Figure 11 shows an example of the results. On the left side, the minor distance was to first image of the standard ASTM E23. This means, that the assessment Charpy test specimen is closely near at the first pattern (ductility 10% see figure 6). On the right side, the minor distance was the seventh (ductility 70%).

Figure 11. Graphical representation of the pattern closest to the tested image using the Euclidean distance and the shape descriptors Area (Left Image) and equivalent diameter (Right Image)

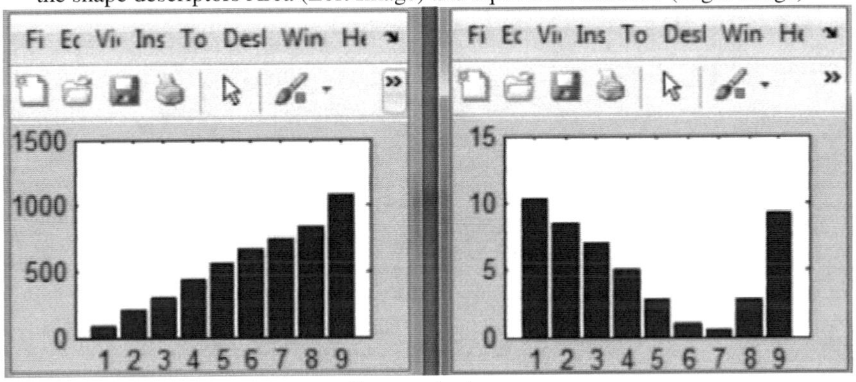

Source: Author

4. Results and validation process

Validation process was made by two experts. They observed at the microscopy the Charpy test specimen images. A summary is shown in tables V and VI. In table I you can see the results of ductility percentage using area as form descriptor.

Table I. Ductility percentage in each of the replicas in the ASI 304 steel using the area as a shape descriptor.

Name specimen	Ductility percentage (%) test 1	Ductility percentage (%) test 2	Ductility percentage (%) test 3	Ductility percentage (%) test 4	Ductility percentage (%) test 5
PRUB304-1	40	40	40	40	40
PRUB304-8	40	40	40	40	40
PRUB304-13	40	40	40	40	40
PRUB304-17	40	40	40	40	40
PRUB304-20	40	40	40	40	40

Source: Author

Table II. Ductility percentage in each of the test in the ASI/SAE 1020 steel using the area as a shape descriptor.

Name specimen	Ductility percentage (%) test 1	Ductility percentage (%) test 2	Ductility percentage (%) test 3	Ductility percentage (%) test 4	Ductility percentage (%) test 5
PRUB304-1	10	10	10	10	10
PRUB304-8	10	10	10	10	10
PRUB304-13	10	10	10	10	10
PRUB304-17	10	10	10	10	10
PRUB304-20	10	10	10	10	10

Source: Author

Table III. Ductility percentage in each of the replicas in the ASI 304 steel using the equivalent diameter as a shape descriptor.

Name specimen	Ductility percentage (%) test 1	Ductility percentage (%) test 2	Ductility percentage (%) test 3	Ductility percentage (%) test 4	Ductility percentage (%) test 5
PRUB304-1	20	20	20	20	20
PRUB304-8	60	60	60	60	60
PRUB304-13	80	80	80	80	80
PRUB304-17	50	50	50	50	50
PRUB304-20	70	70	70	70	70

Source: Author

Table IV. Ductility percentage in each of the test in the ASI/SAE 1020 steel using the equivalent diameter as a shape descriptor.

Name specimen	Ductility percentage (%) test 1	Ductility percentage (%) test 2	Ductility percentage (%) test 3	Ductility percentage (%) test 4	Ductility percentage (%) test 5
PRUB304-1	70	70	70	70	70
PRUB304-8	80	80	80	80	80
PRUB304-13	80	80	80	80	80
PRUB304-17	70	70	70	70	70
PRUB304-20	40	40	40	40	40

Source: Author

Tables III and IV show that the percentage of ductility obtained by the algorithm is variable depending on the specimen. This indicates that being the same type of material, different results are observed. However, with the area (Tables I and II) as a descriptor the results are very uniform.

Table V. Comparison of results between the Algorithm and the experts in the area of materials in AISI 304 steel at room temperature.

Name specimen	Ductility /Brittleness percentage (%) obtained by the algoritm(descriptor area)	Ductility /Brittleness percentage (%) obtained by the expert 1	Ductility /Brittleness percentage (%) obtained by the expert 2
PRUB304-1	40/60	40/60	40/60
PRUB304-2	40/60	40/60	40/60
PRUB304-3	40/60	40/60	40/60
PRUB304-4	40/60	40/60	40/60
PRUB304-5	40/60	40/60	40/60
PRUB304-6	40/60	40/60	40/60
PRUB304-7	40/60	40/60	40/60
PRUB304-8	40/60	40/60	40/60
PRUB304-9	40/60	45/55	45/55
PRUB304-10	40/60	35/65	35/65

Source: Author

Table VI. Comparison of results between the Algorithm and the experts in the area of materials in AISI/SAE 1020 steel at room temperature.

Name specimen	Ductility /Brittleness percentage (%) obtained by the algoritm(descriptor area)	Ductility /Brittleness percentage (%) obtained by the expert 1	Ductility /Brittleness percentage (%) obtained by the expert 2
PRUB304-1	10/90	5/95	5/95
PRUB304-2	10/90	10/90	10/90
PRUB304-3	10/90	10/90	10/90
PRUB304-4	10/90	10/90	10/90
PRUB304-5	10/90	10/90	10/90
PRUB304-6	10/90	10/90	10/90
PRUB304-7	10/90	10/90	10/90
PRUB304-8	10/90	10/90	10/90
PRUB304-9	10/90	15/85	15/85
PRUB304-10	10/90	10/90	10/90

In tables V and VI, the error observed was 20%.

5. Conclusion

The application of graphite in the ASI 304 specimens turned out to be an alternative determinant to carry out the image processing, allowing the software to differentiate the areas formed in the test and thus subsequently go on to make a numerical comparison with the areas formed and the Pattern image areas using Euclidean distance. Evaluation by expert in materials was the determining factor for determining the degree of reliability of the algorithm since it allowed to know with certainty which descriptor of form (Equivalent area or diameter) was ideal for carrying out this type of image processing. A very low error rate (20%) has been presented when area is used as a shape descriptor. However, it is valid clarify that the percentage of error is supported by the application of ranges discrete that the ASTM E23 standard has. Likewise, it was evidenced that the equivalent diameter is not the most adequate descriptor for this type of application, since the results were not in agreement with the results of the experts. External factors at the time of image capture is a critical factor since any anomaly that occurs at the time of capturing the image will affect the digital processing of images for this it is necessary to eliminate all the noises or factors outsiders involved in the scene An open problem is still precision, consistency and efficiency of the algorithm. Future work will be focused on the development of a sophisticated method of deriving an optimized high-precision matching result under the influence of noise and illumination sources.

6. References

[1] P. J. Bagga, M. A. Makhesana, K. Patel, and K. M. Patel, "Tool wear monitoring in turning using image processing techniques," Mater. Today Proc., vol. 44, pp. 771–775, 2021.

[2] N. Dey, J. Chaki, L. Moraru, S. Fong, and X.-S. Yang, "Firefly Algorithm and Its Variants in Digital Image Processing: A Comprehensive Review," in Applications of Firefly Algorithm and its Variants: Case Studies and New Developments, N. Dey, Ed. Singapore: Springer Singapore, 2020, pp. 1–28.

[3] E. R. Dougherty, Digital image processing methods. CRC Press, 2020.

[4] G. K. Ijemaru et al., "Image processing system using MATLAB-based analytics," Bull. Electr. Eng. Informatics, vol. 10, no. 5, pp. 2566–2577, 2021.

[5] I. R. Fermo, T. S. Cavali, L. Bonfim-Rocha, C. L. Srutkoske, F. C. Flores, and C. M. G. Andrade, "Development of a low-cost digital image processing system for oranges selection using hopfield networks," Food Bioprod. Process., vol. 125, pp. 181–192, 2021.

[6] P. Salehi and N. Behzadfar, "Investigation and simulation of different medical image processing algorithms to improve image quality using simulink matlab," Signal Process. Renew. Energy, vol. 5, no. 4, pp. 15–28, 2021.

[7] O. Cantó-Navés, X. Marimon, M. Ferrer, and J. Cabratosa-Termes, "Comparison between experimental digital image processing and numerical methods for stress analysis in dental implants with different restorative materials," J. Mech. Behav. Biomed. Mater., vol. 113, p. 104092, 2021.

[8] S. Mahankali and G. Valikala, "Comparison of Compressive Strength of M30 Grade Concrete with Destructive and Nondestructive Procedures Using Digital Image Processing as a Technique," Adv. Civ. Eng., vol. 2022, 2022.

[9] B. E. Romero-Tarazona, C. L. Rodriguez-Sandoval, J. G. Villabonai-Ascanio, and A. D. RincónQuintero, "Development of an artificial vision system that allows non-destructive testing on flat concrete slabs for surface crack detection by processing of digital images in MATLAB," IOP Conf. Ser. Mater. Sci. Eng., vol. 844, no. 1, 2020, doi: 10.1088/1757-899X/844/1/012058.

[10] Y. Yang, H. Wang, Y. Yang, and H. Zhang, "Evaluation of the evolution of the structure of cold recycled mixture subjected to wheel tracking using digital image processing," Constr. Build. Mater., vol. 304, p. 124680, 2021.

[11] A. Surendranath and P. V Ramana, "Recycled materials execution through digital image processing," Mater. Today Proc., vol. 46, pp. 8795–8801, 2021.

[12] D.-H. Xia et al., "Review-material degradation assessed by digital image processing: Fundamentals, progresses, and challenges," J. Mater. Sci. Technol., vol. 53, pp. 146–162, 2020, doi: https://doi.org/10.1016/j.jmst.2020.04.033.

[13] V. S. Barbosa, L. A. C. de Godois, K. E. Bianchi, and C. Ruggieri, "Charpy impact energy correlation with fracture toughness for low alloy structural steel welds," Theor. Appl. Fract. Mech., vol. 113, p. 102934, 2021.

[14] C. Wang et al., "A method for directly measuring fracture toughness and determining reference temperature for RPV steels by Charpy impact test," Eng. Fract. Mech., vol. 243, p. 107526, 2021.

[15] D. R. Hetrick, S. H. R. Sanei, O. Ashour, and C. E. Bakis, "Charpy impact energy absorption of 3D printed continuous Kevlar reinforced composites," J. Compos. Mater., vol. 55, no. 12, pp. 1705–1713, 2021.

[16] C. Li and Z. Zhou, "Charpy impact behavior of a novel stainless steel powder wire mesh composite porous plate," Materials (Basel)., vol. 14, no. 11, p. 2924, 2021.

[17] R. Chaouadi and R. Gerard, "Development of a method for extracting fracture toughness from instrumented Charpy impact tests in the ductile and transition regimes," Theor. Appl. Fract. Mech., vol. 115, p. 103080, 2021.

[18] ASTM, Standard Test Methods for Notched Bar Impact Testing of Metallic Materials. 2018.

[19] J. R. M. Vilet, "Procesamiento digital de imágenes," Fac. Ing. UASLP, Doc. PDF, Dispon. en la página!` http//read. Pud. com/downloads159/ebook/711796/Procesamien to Digit. Imagenes. pdf?`[Citado 22 septiembre 2012], 2005.

[20] N. L. S. Palomino and U. N. R. Concha, "Técnicas de segmentación en procesamiento digital de imágenes," Rev. Investig. Sist. e Informática, vol. 6, no. 2, pp. 9–16, 2009.

[21] L. Lozano and J. Fernández, "Razonamiento basado en casos: Una visión general," Recuper. el, 2008.

[22] C. L. Sandoval Rodriguez, A. A. Barros, and S. Herreño, "Automatic Classification of mechanical vibration patterns in rotating machinery affected by unbalance," 2021.

[23] S. E. Rigdon, R. Pan, D. C. Montgomery, and C. M. Borror, Design of Experiments for Reliability Achievement. John Wiley \& Sons, 2022.

Educational strategies applied to the logistics system of transport, treatment and final disposal of used lubricating oils in Colombia

J C Murcia, L J Hernández Corporación Universitaria Minuto de Dios – UNIMINUTO, Maestría en educación, Ingeniería informática, Bogotá, Colombia y D A López Universidad Distrital Francisco José de Caldas, ingeniería Eléctrica Bogotá, Colombia

jennifer.murcia@uniminuto.edu, leydy.hernandez-v@uniminuto.edu.co, dalopezs@udistrital.edu.co

Abstract. In Colombia, one of the problems that is evident and that is gaining strength is related to the inadequate management of dangerous resources such as used lubricating oils (ALU) from automobiles and motorcycles, where it is observed that there is no adequate pedagogical strategy on logistics systems or logistics processes that allow optimal final disposal, thus avoiding contamination and alteration in the environment. The methodology that was used for this investigation is of a mixed type, with a descriptive investigation, where the ignorance of the conditions for the adequate handling, treatment and disposal of used lubricating oils can be identified, emphasizing small and medium-sized mechanical companies. Mechanical workshops mainly in neighborhoods located in the city of Bogotá and mostly in locations that do not have the latest technology to provide such service. As results of the investigation, it was found that there is no educational model of socialization of a structured logistics system for establishments that provide mechanical services, in addition to the lack of training and awareness on the part of companies and entities dedicated to the use of this type of substances, denoting the importance of providing learning tools from the proposed educational strategy.

1. Introduction

One of the main most notable problems are those related to environmental deterioration caused by multiple factors derived from human behavior and processes, the indiscriminate use and disposal of used lubricating oils - ALU, which generate a considerable impact considering that their use It is essential around the planet, its bad disposition brings with it contamination and alteration of the environment in which it is compromised.

Currently, there is ignorance of the conditions and necessary elements, for an adequate management and disposal of the ALU, by small and large establishments in which this hazardous waste is handled. This is due to the lack of training and awareness on the part of companies and entities dedicated to the development and use of this type of substance.

According to [1], ALUs are considered hazardous waste due to their flammability and toxicity characteristics. They are subject to a management plan in accordance with what is indicated in the General Law for the Prevention and Comprehensive Management of Waste (LGPGIR), improper handling of these can cause contamination in water, soil and air.

The purpose of the research work is to contribute to the creation of transport routes that optimize the collection times of the ALUs, deliver them to the collection centers so that they can carry out their process so that they can be reused with the in order to avoid future environmental problems. In addition to sensitizing the owners of motorcycle workshops, especially so that they carry out the initial process of collecting the waste in an adequate and optimal way.

2. Related Works (Literature)

Using various search sources, the research and / or articles that have been published nationally and internationally have been reviewed. The most relevant characteristics are described for each of them.

In accordance with the above, when reviewing the literature on the subject, some works from an international scope stand out, such as: according to the company's research source (Cempre Uruguay), there are numerous processes to recycle hazardous waste (oils) that affect the environment. environment and well-being in terms of the health of an entire community in general, with slight cleaning processes. The most common process used for many years, and in particular for small or medium-scale applications, is the acid-clay process [2].

In this article published by the same company and according to official data, currently in Spain 77% of the 220,000 tons of the total is collected. The remaining 23% (50,000 tons) is out of control. Of the total used oils collected each year, 66,000 tons go to incineration or combustion, 51,000 tons go to cogeneration, 49,000 tons go to regeneration and 4,000 tons go to other destinations.

In Uruguay, the only form of safe disposal for the ALU is the burning in Portland cement kilns, because the high combustion temperatures (over 1000ºC) and the long residence times ensure the destruction of potentially dangerous chemical substances that they may be present.

IOP Conf. Series: Materials Science and Engineering 1299 (2024) 012007 doi:10.1088/1757-899X/1299/1/012007

The authors [3] of the University of Zaragoza, Spain, wrote an article called "Public Policies for the Promotion of Recycling: The Regeneration of Used Oils", in order to investigate which process is the most suitable for the adequate reuse of ALUs. . This project analyzed the effectiveness of public policies against this great environmental problem. Also, the operation of the market for these oils in Spain was analyzed, in order to design the structure of public policies that support the regeneration of these residual oils.

According to the (International Center for Environmental Law Studies) CIEDA-CIEMAT of Spain. Order APM / 205/2018, of February 22, establishes the criteria to determine when processed used oil from ALU treatment for use as fuel ceases to be waste pursuant to Law 22/2011, of July 28 , waste and contaminated soil. Directive 2008/98 / EC of the European Parliament and of the Council of 19 November 2008 introduced a new procedure to define the criteria by which it can be established that a certain flow of waste can no longer be considered waste. On the other hand, in accordance with the provisions of article 8 of Law 22/2011, of July 28, on contaminated waste and soil, which regulates the principle of waste hierarchy, used oils must be used as a first priority for the regeneration (recycling) and as a second to energy recovery, either directly or by obtaining fuels.

Taking into account what was mentioned by [4], where they indicate that the Argentine market for lubricating oils for automotive use is dominated by five large regional and international companies, which dedicate large resources and marketing actions to the sale of high-value oils. aggregate. Despite the existence of adequate and easily accessible technologies for the reprocessing of used oil, and that the final product obtained achieves excellent quality standards, no company has been developed in Argentina that produces and markets reprocessed oils on a large scale. Nor does Argentine law establish obligations on petrochemical companies for the recovery of used oil. In other countries, producers must recover high percentages of the total volume sold. For example, in Brazil it reaches 40% of total sales.

In the recommendations made in his thesis, the student Manuel Rimapa Ruiz of the National University of the Peruvian Amazon, (2011), announces that every company that markets used oils must provide constant training to its workers in the handling of oils. used lubricants, in coordination with the sector authorities.

According to what is indicated in the thesis of [5], industrial growth on a continuous basis leads to a significant increase in the demand for natural resources and their derivatives. Lubricating oils are no stranger to this growing demand; growth in industrial development is unthinkable. Such a situation, opens the mind to visualize the problems that are generated around the world, due to not making a correct storage of used oils within the workshops.

From a national level, [6] those who speak in the general context of residual oils stand out about a technical manual for the management of used lubricating oil from the Ministry of the Environment, which was implemented in order to design a plan of activities, workshops and Socialization of a teaching booklet that promotes awareness and training on the proper handling of used lubricating oil for the prevention of pollution and care for the environment. This occurred in the city of Florencia (Caquetá) which increased its automotive fleet in recent years, especially motorcycles, going from 9,987 motorcycles registered in 2012 to 44,300 as of March 31, 2018 [7], which indicates a greater volume of hazardous waste generated in establishments that carry out motorcycle maintenance, especially used lubricating oil.

According to [8], used oils of automotive and industrial origin are classified as hazardous waste from mineral oils, not suitable for the use for which they were originally intended, or mixtures and emulsions of oil and water wastes or hydrocarbons and water. Their dangerous characteristics vary according to the processes or equipment in which they have been used. For this reason, Colombia has the challenge of giving comprehensive management to used oils, which guarantees their environmentally safe management in order to generate a national strategy within the applicable legal framework that guides the management of used oils. For this, they propose a technical manual that allows knowing how to implement the proper handling of used lubricating oils of automotive and industrial origin, published in 2006 by the Ministry of Environment, Housing and Territorial Development, today the Ministry of Environment and Sustainable Development.

On the other hand, the authors [9], in their research article, speak of responsible management that is required for the proper use of residual motor oils, which are a highly polluting material, since they can cause damage to the environment when they are dumped on the ground or in water currents including sewers. This can result in contamination of groundwater and soil. The used lubricating oil contains various chemical compounds such as heavy metals, (for example, chromium, cadmium, arsenic, lead, among others), polynuclear aromatic hydrocarbons, benzene and sometimes chlorinated solvents, PCBS, etc. These chemical compounds have a direct effect on human health and several of these products are carcinogenic. A universal concern is the supply of water; used lubricating oils penetrate the ground and automatically contaminate surface and ground water. When they are eliminated in city sewage systems, they end up in wastewater treatment plants, but currently, because there are very few of these plants in Colombia, the vast majority of these oils end up in the environment and in the surface water.

In the Thesis [10] in summary, they make it known that for the development of their thesis they carried out an initial situational diagnosis in Florence (Caquetá) and identified that currently the normative technical management of this type of waste is not applied in the municipality, for what each generator, individually, performs the management of used oils with evident flaws, both in storage and transport and in the final disposal, which is confirmed in the main diagnosis in which the survey was applied to the collectors primary. For the development of the proposal, the current legislation and regulations on the management and disposal of used lubricating oils were analyzed. Likewise, the guidelines and procedures established by control agencies and environmental authorities for the management and disposal of used lubricating oils applicable in each link of the chain were analyzed.

A Reverse Logistics Network (RLI) has been proposed for the city of Pereira [11], where the waste generating points (ALU), distances, associated logistics costs and the type of collection vehicle are taken into account; the system model was modeled and simulated in MATLAB as a capacity constrained vehicle routing problem (CVRP - Capacited Vehicle Routing Problem), the solution points to the order of visit of vehicles to the generating points and was carried out in two steps : savings and cluster heuristics.

A web-based decision system [12] allows the establishment of calendars, milestones and follow-ups based on consumer demand, which in this case is an ALU re-refining plant, the inventories level of the plant closes the link of the RLI. The logistics process has four basic characteristics: the reasons for the return, the type of product, the recovery process and the actors involved. Other aspects that were taken into account were: storage capacities of the sources of the waste, location, characteristics of the containers, heterogeneity of the collection vehicles, quality of the ALU labeled as reusability, the plant's production plans (MRP - Material Requirements Planning and MPS - Master Scheduling Planing).

Characterization of the management process of used lubricating oils and main transport logistics systems.

A. Management of used lubricating oils

To understand the problem further, the difference between vehicle and motorcycle lubricating oil must be taken into account, which do not have the same characteristics as shown below, according to [13] "the greatest obstacle to the exchange of lubricants between cars and motorcycles is presented by the clutch, this because it tends to slide on a motorcycle. Therefore the materials with which they are manufactured are not compatible, and specifically vary in their "anti-friction" additives in addition to being incompatible by contact between the parts, their content is based on different components taking into account that the time interval between changes it is completely different since motorcycle lubricant is usually carried to the extreme more continuously, this generates accelerated wear of its components unlike that used in a car, for this reason it must be replaced more often and There must be a regulated control of the ALU of motorcycle-type vehicles, hence the focus of the project.

According to [9], "all used oils are regenerable, although in practice the process and its cost make this alternative unfeasible" recycling this type of waste is viable because they can be treated in their entirety, On the other hand, they state that "Regarding alternatives for the use of used motor oils, reuse is highlighted as long as its characteristics allow it. This reuse begins with a pretreatment that includes removal of contaminants, insolubles, water, sediments and oxidation products, through the use of different physicochemical processes such as its heating, filtration, dehydration and centrifugation. Regeneration through different treatments is possible, recovering the base lubricating material present in the original oil, so that it can be reprocessed for its use." These can be integrated back to a useful life after a purification process, the oils collected in the The proposed route has great reprocessing capacity because the training will provide the information and advice necessary for the optimal disposal of this type of ALU.

Table I. Characteristics of the (ALU).

Characteristics	Automotive
Viscosity at 40 oC, SSU	97-120
Gravity at 15.6 oC, oAPI	19-22
Specific gravity at 15.6 oC	0.9396_0.8692
Water,% vol.	0.2-33.8
Sediments,% vol.	0.1-4.2
Insoluble in benzene, wt%	0.56-33.3
Soluble in gasoline,% vol.	2.0-9.7
Flash point, oC	78-220
Heat output, MJ / Kg	31.560-44.880

Source: Recovered from "Technical manual for the management of used lubricating oils of automotive and industrial origin, by the Ministry of Environment and sustainable development 2014. Physicochemical characteristics of used oils. P. 18.

Table II. Allowable pollutant levels in (ALU).

Substance	Maximum allowable concentration (mg / Kg / -ppm)
Polychlorinated biphenyl (PCBs)	50
Total organic halogens (as Chlorine)	1000
Arsenic	-
Cadmium	-
Chrome	10
Lead	100
Sulfur	1.7% weight

Source: Recovered from "by which the cases in which the combustion of waste oils is allowed and the technical conditions to carry it out are established", by the Ministry of Environment and sustainable development. 1998. Table of percentage for mixing and maximum limits of contaminants in used oils. P. 6

For the environmental protection agency (EPA), which is an agency of the United States federal government in charge of protecting human health and protecting the environment: air, water and soil, used oil must have a clear definition, these indicate which is one that has been refined from crude oil or any synthetic oil that has been used and this use is contaminated with physical and chemical impurities [14]. The impurities that can be ingested in the oil makes it no longer have the initial properties, impurities such as: earth, water, metal particles and chemical products that in one way or another affect the effectiveness of the oils, that's why When it comes to car or motorcycle oils, they must be changed or replaced in a range of miles traveled by these vehicles, this type of oil is known as lubricant.

According to what the author [15] indicates in his thesis, lubricating oil is solid, semi-solid or liquid substances of animal, vegetable, mineral or synthetic origin, which can be used to reduce friction between moving parts and mechanisms.

Within several investigations, it is evident that the function of lubricating oils is not only limited to reducing friction between materials, but also fulfills other functions that are important for achieving effective functionality within the activities it performs, within these they can be indicated [16]: coolant, impurities remover, sealant, anticorrosive and antiwear, and energy transmitter.

B. Logistics transport system.

In the presentation of [17] he discloses the appropriate procedure for the transportation of ALUs, which is divided into three phases:

Phase 1: Packaging

There are 3 options for the transportation of 55-gallon capacity drums, 55-gallon capacity containers and tank trucks. The containers must be in perfect condition, without dents or damage to their edges, be made of materials resistant to corrosion and that can be hermetically closed before starting their movement, avoiding at all times the spillage of used lubricating oil. To fill the 55 Gallon drums, a 10 cm free edge must be left.

Phase 2: Labeling

Each tank, drum or storage system must be labeled with the words "Liquid substance potentially dangerous for the environment-it contains used lubricating oil", the vehicle must have on a plate the United Nations number corresponding to the character of the product transported (UN 3082), on all visible faces of the unit and the front of the cargo transport vehicle cabin.

Phase 3: Transportation

Vehicle conditions: in the case of a vehicle with 55-gallon drums or tanks, these must be fixed to the vehicle through the use of fastening devices used especially for this purpose. If the mobilization is carried out in a tank car, the length of the chassis must protrude from the rear end of the tank, the tank must have a plate with the name of the manufacturer, the standard or building code, the date of manufacture, capacity and number of compartments. . The tank, drums or storage system must be resistant to hydrocarbon action to ensure complete confinement of the used lubricating oil. There must be a communication system (cell phone, radiotelephone, radio or other) and its respective license issued by the competent authority for applicable cases.

Elements: Pump for loading or unloading used lubricating oils; hose for loading or unloading used lubricating oils; fire extinguishers; oleophilic material; personal protection elements (Overalls, boots, gloves resistant to the action of hydrocarbons).

Procedure for loading ALU in the facilities of a collector or warehouse:

Vehicle inspection: The vehicle driver must check the integrity of the equipment daily prior to the first trip and must make sure to carry a copy of the following safety documents: Safety sheet of used lubricating oils, presented in Annex 3 of this Manual, Contingency Plan for consultation in case of emergency.

Location of the vehicle: Once in the facilities of the Collector or Warehouse, the vehicle must be located in the temporary storage area where it does not cause interference, in such a way that it is in a quick exit position, according to instructions given by the person in charge of the installations.

Preparing for Pumping: Inspection, the vehicle driver and facility manager should visually inspect the tanks, storage area, and valves to verify their condition and identify the presence of leaks or leaks of used lubricating oil.

In informal workshops, there is no system that allows them to have efficient control of both their entrances and their exits, they can implement a system that provides them with a competitive, proactive and innovative differential, compared to other workshops. Accessing the digital age as indicated in the article from the Uniminuto inclusion and development magazine in its publication of December 10, 2017 in which it reveals the challenges Colombia faces taking into account the different digital changes.

Taking into account the article in the academic journal [18], Characterization of the actions for the separation of waste at the source of the Caracolí, Potosí and Tres esquinas neighborhoods of Ciudad Bolívar, the importance of the classification is determined. and correct disposal of waste, because it has a negative impact on exposed environments, if the high level of waste generated from various activities is taken into account and projecting it into the future, it is concluded that it is vitally important to awareness and treatment of waste at its points of origin.

According to the reasoning that has been carried out regarding the quality of employment that is being generated in the city of Bogotá, the sectors with the highest employability indicator are also those that have a high index of informal work, given that they lack income decent, seasonal work, casual or fixed-term employment. [19] states that:

This article presents an investigation carried out in the center of the city of Santiago de Cali, which addresses one of the most complex and growing realities of recent years at the national level: informal work. Through four life stories, the human social drama that informal workers live daily, for not having real job opportunities, for not having adequate educational levels, for lack of economic resources and for the evident disinterest of the State is presented.

3. Methodology

The research project, which is underway, will be prepared under a quantitative and qualitative methodological approach, based on the gathering of information through surveys, interviews and photographic records, in addition to the analysis and observation that has been made through visits made to workshops located in specific areas; This is a descriptive and explanatory research, becoming a mixed research based on the book "Research Methodology" of [20].

Starting from the descriptive point of view, with the aim of determining the behaviors they present regarding the handling of ALUs in motorcycle workshops in the Santander, San Victorino and Villa Nelly neighborhoods; Since this type of research seeks to clarify the behaviors that are the object of study, such as the number of approximate oil changes and amounts collected daily in order to determine the volume of waste generated, in addition to identifying behaviors and disposal of this type of substances; on the other hand, the type of explanatory research in view of the fact that in the course and progress of the project, responses will be given to the factors and causes that lead to the misuse of ALUs, this in order to provide an alternative and raise awareness regarding the problems studied according to [21].

Following the designed methodology, the identification and recognition of the environmental aspects and impacts generated by the inadequate management of ALUs in motorcycle workshops in the previously described areas will be carried out.

According to this, it was evident that most of the people who make the oil change in the Santander, San Victorino and Villa Nelly neighborhoods, are not clear about the process that these oils receive after being collected from their workshops and in some cases, these used oil residues - ALUs are collected by untrained people, mismanaging this residue.

On the other hand, it was analyzed that in these neighborhoods there is no appropriate route for the collection of used oil, which shows that there is no optimal route to take them where they are given proper management of this waste and the capacity and the types of containers or basins that are used to collect these used oils. This allows us to broaden the vision of this problem a little more and rectify the importance of implementing a transportation system in order to minimize time and optimize resources, such as project design (ALU), giving possible solutions that contribute to it. time with caring for the environment.

Population and sample are based on:
In Bogotá DC, the capital of Colombia, there are a large number of establishments dedicated to guaranteeing the useful life of motorcycle-type vehicles. If the size of the city is taken into account, it is complex to try to tackle the entire problem, which focuses on the damages caused by the ALUs derived from the bad disposition and management by those involved, due to this situation, three critical and strategic points of the city are identified, the Santander, San Victorino and Villa Nelly neighborhoods in which it is intended to address and identify behaviors and practices associated with the problems, the result of poor management by the locals and people who work there.

The number of premises for investigation varies in each area, approaching with a range of (80 to 100) establishments in each neighborhood, from which it is taken (10%) to carry out data collection applying in total (30) interviews in the previously described areas, with this we proceed to collect and analyze information for a subsequent interpretation of results; The interviewees proceed to apply the selected method for the acquisition of information, "surveys" that will be applied randomly to a number of people according to [22] accessible population, since they work in the premises and represent a subset of the population temporarily Available for the study, these are in the delimitation of the described areas, both owners and employees who possess basic or advanced knowledge about the practices and procedures applied are involved, in addition to the amounts of ALU waste derived from the activities carried out in a continues in establishments.

4. Results

The main objective of the research seeks to propose a design of a logistics transport system for lubricating oils used in motorcycle workshops.

The objectives that allow achieving this main objective are founded on as specific objectives and are formed in the phases of the development of the investigation, beginning with the collection of information from the main transport logistics systems, for the proper selection of the best, with the reason for transporting the lubricating oil to have a proper handling and final disposal thereof. Next, it is important to identify the volume of waste originating or derived from lubricating oils from motorcycle maintenance. Next, we seek to define optimal transport routes in order to minimize collection times, improving resources. To end with the socialization of the strategies derived from the transport logistics system and the proper handling of lubricating oil in motorcycle workshops.

The investigation is in the first two phases:

Phase 1: Database - Collection of information (State of the art).

Phase 2: Information gathering and data analysis.

This shows that the investigation is ongoing, in the database and information gathering phases, for which it was carried out in phase 1, the analysis of the information found, where it was possible to identify and determine that there is no Evidence within the literature of route designs for the transportation of used lubricating oils - ALU, which shows that there is a knowledge gap in this regard and that it is relevant to carry out the research.

In phase 2, within the collection of information, a technical instrument for data collection (Survey) was designed, which was validated by experts in the field to assess knowledge about the proper management of used lubricating oils by the owners of establishments such as motorcycle workshops. This phase is still ongoing, to complete its effective development, applying the instruments.

5. Conclusion

For the ongoing investigation, the phases developed and which are still being executed, it can be preliminarily concluded that there is no transport logistics design supported by the proper process of handling and final disposal of used lubricating oils, so It can be seen in the literature that, although there are works that use the handling of oils, the final disposal, this has not been worked from the point of view of the motorcycle sizes sector, thinking of the small businessman.

In the partial application of the technical research instrument (survey) to motorcycle workshops, a lack of knowledge and proper handling of lubricating oils was observed, since from the empirical knowledge of many mechanics they carry out the work without any care or recognizing that this It is a hazardous waste, found oil spills, dirt, little training and training to do the job of changing motorcycles oil in a more suitable and proper way.

References

[1] Jurado, A. (2016, 4 de agosto). Contaminación y manejo de aceites lubricantes usados. Hoy en la Salle. Recuperado de: http://hoy.lasalle.mx/contaminacion-y-manejo-de-aceites-lubricantes-usados/

[2] Cempre Uruguay, (2016-2017). Aceites usados. Uruguay. Cempre Uruguay. Recuperado de: http://www.cempre.org.uy/index.php?option=com_content&view=article&id=79&Itemid=85

[3] Arner, A; Barberán, R; Mur, J, (2003). Las políticas públicas de Fomento del Reciclaje: La Regeneración de Aceites Usados, Revista de economía pública, 167, (4): 33-55.

[4] Nervo, Gustavo; Sanna, Domingo (2011). REPROCESO Y COMERCIALIZACIÓN DE ACEITE LUBRICANTE USADO, Recuperado: https://ucema.edu.ar/posgrado-download/tesinas2012/Tesina_MBA_UCEMA_Montero.pdf
[5] Masías Puma, R., Pichuca Chanqueti, E. D., & Pariona Gutiérrez, R. A. (2017). Implementación del plan y manejo de reciclaje de lubricantes en el taller mecánico de motos Ssenda para reducir la contaminación ambiental en el distrito de Ate Vitarte año 2017.
[6] Restrepo, Lina; Rojas, Ángela (2016). Diagnóstico y programa de capacitación del aceite usado - sector mantenimiento motocicletas Florencia Caquetá. Recuperado de: https://core.ac.uk/download/pdf/143464438.pdf
[7] Registro Único Nacional de Tránsito. (2018). 52% de las motos matriculadas en Bogotá son de cilindraje mayor a 125 centímetros cúbicos (002). Recuperado de: http://www.runt.com.co/sites/default/files/Bolet%C3%ADn%20002%20de%202018.pdf
[8] Ministerio de Ambiente y Desarrollo Sostenible. (2014). Manual técnico para el manejo de aceites lubricados usados de origen automotor e industrial. Bogotá D.C, Colombia Ministerio de Ambiente y Desarrollo Sostenible. Recuperado de: https://acp.com.co/web2017/images/pdf/combustiblesylubricantes/FAU/Manual-AU-final-14.pdf.
[9] Silva W. F., Quiñones B. E. & Tejada T. C., (2017). Caracterización físico-química de aceites usados de motores para su reciclaje. Cartagena, Colombia. Prospectiva Vol. 15 Recuperado de http://www.scielo.org.co/pdf/prosp/v15n2/1692-8261-prosp-15-02-00135.pdf
[10] L. Rodríguez, «Tesis Maestría - Programa : lineamientos para el manejo y disposición final de aceites lubricantes usados en Florencia Caquetá,» 28 Julio 2011. [En línea]. Available: https://repository.javeriana.edu.co/bitstream/handle/10554/6082/tesis125.pdf?sequence=1&isAllowed=y.
[11] Castañeda Jiménez, J., & Cardona Arias, J. A. (1 de Junio de 2017). Diseño de una Red de logística Inversa para recolectar Aceite Vehicular Usado en la ciudad de Pereira implementando CVRP. (H. A. Gallego Becerra, Ed.) Scientia et Technica, 22(2), 150-160. doi:10.22517/23447214.9243
[12] Repoussis, P., Paraskevopoulos, D., Zobolas, G., Tarantilis, C., & Ioannou, G. (2009, Junio 16). A web-based decision support system for waste lube oils collection and recycling. European Journal of Operational Research, 195(3), 676-700. doi: 10.1016/j.ejor.2007.11.004
[13] Total, Blog. (14 de junio 2017). Diferencias entre lubricantes para moto y para coche. Tu Experto en Lubricantes. Recuperado de: http://blog.total.es/diferencias-lubricantes-moto-y-coche/
[14] Agencia de Protección Ambiental de Estados Unidos. (2017). Definición de Aceite Usado. Washington D. C., Estados Unidos: Agencia de Protección Ambiental (EPA).
[15] Llanos, F. (2017, Septiembre 02). Propuesta para el manejo del aceite usado de vehiculos automotores en el cantón SigSig. Retrieved from Universidad Politécnica Salesiana: file:///D:/Proyecto%202020/UPS-CT002737.pdf
[16] Manual Técnico Aceite Gulf. (2016). Manual Técnico. Argentina: Gulf. Retrieved from http://remex.com.ar/descargas_web/1-1%20Gulf%20-%20Manual%20Tecnico%20de%20Lubricantes.pdf
[17] Giraldo P. K., (2016), Transporte de Aceites Usados. Prezi. Recuperado de: https://prezi.com/vknoewfo6dmz/transporte-de-aceites-usados/
[18] J. Santos, J. Niño y L. Vargas, «Caracterización de las acciones para la separación de los residuos en la fuente de los barrios Caracolí, Potosí y Tres esquinas de la localidad Ciudad Bolívar,» Revista Inclusión & Desarrollo, vol. 4, nº 2, pp. 94-101, 2017.
[19] F. Vazquez, L. Alvarez y A. Mera, «EL TRABAJO INFORMAL EN COLOMBIA: DRAMA SOCIAL HUMANO,» Revista Inclusión y Desarrollo, vol. 3, nº 3, pp. 52-61, 2015.
[20] Bernal, C., (2010). Metodología de la Investigación. Bogotá D.C, Colombia. Pearson Educación.
[21] C. Cabanzo, «Informe Técnico de investigación: Observatorio de educación, infancias y políticas públicas educativas.,» UNIMINUTO, Bogotá, 2017.
[22] Gil, P. S. (2011). Población de estudio y muestra. Obtenido de PDF: http://udocente.sespa.princast.es/documentos/metodologia_Investigacion/Presentaciones/4_%20poblacion%26muestra.pdf

Evaluation of the anaerobic digestion process for low and micro-scale biogas production: a review

N. Y. Castillo-Leon[1], B. E. Tarazona-Romero[1 2], M. E. Carvajal Gamarra, O. A. Acosta-Cardenas[1], Y. P. Quizena-Fernandez and M. Y. Alarcon-Guerrero[1]

[1] Automation and Control Energy Systems Research Group (GISEAC), Faculty of Natural Sciences and Engineering, Electromechanical engineering, Unidades Tecnologicas de Santander (UTS), Student Street No 9-82, 680005, Bucaramanga, Colombia.
[2] Energy in Building research group (ENEDI), Doctoral Program in Energy Efficiency and Sustainability in Engineering and Architecture, Department of Machines and Thermal Engines, University of the Basque Country (UPV / EHU), Engineer Torres Quevedo Square, 1, 48013, Bilbao, BI, Spain.

E-mail: nycastillo@correo.uts.edu.co

Abstract. A theoretical analysis of biogas production technologies classified in two subgroups; low and micro scale, was carried out using the methodology of formal concept analysis (FCA) together with the VOSviewer and Conception Explorer tools for information processing, a series of attributes that each subsystem must comply with in terms of the sustainability of each technology were analyzed. The information selected for the development of the analysis required a quality study, based on the percentile of the journals in which the manuscript was published. The search for information was carried out in web sites such as: Scopus, Web Of Science, Dimensions and Lens among others. In total, 105 sources of academic information were analyzed, selecting 56 articles and 18 book chapters for the development of the work. In conclusion, it can be affirmed that low-scale biogas plants present more favorable characteristics than micro-scale plants; however, both technologies have a high potential for application worldwide in isolated or rural areas, supplying electric or thermal energy to less favored communities. Finally, it is important to point out that the methodology developed has proven to be a relevant tool for the classification of information and serves as a first step to group and relate search concepts by means of a deterministic set of attributes.

Keywords: Biodigester, Digestion anaerobia, Formal Analysis of Concepts, Production de biogas, Renewable Energy

1. Introduction

The world is currently undergoing an energy crisis due to the continuous exploitation and use of fossil fuels, which is why different renewable energy sources such as solar, wind and biomass have been explored [1]. Additionally, the exploration of this type of energy has been an urgent global mission due to the fact that fossil fuel reserves are limited and their accelerated use has negatively impacted human health and climate change [2]. Understanding the above, biomass resources are an attractive alternative with high potential for renewable energy production. Biomass, including plant and waste generated by animal and human activities, is a renewable organic source that can be reused [3][4].

Globally, digesters have been planned, designed, constructed and maintained, incurring active applications and technological developments of low and micro-scale biogas [5]. Currently, there are a large number of unreported biodigesters, however, indicators relate that there are more than 4000 small-scale (decentralized) digesters in Germany, 350 in Austria, 72 in Switzerland and 65 in the United Kingdom, related to Europe alone [6] [5]. In Asia, China has about 30 million biodigesters, India 3.8 million, Nepal 0.2 million and Bangladesh about 60,000. In the United States and Canada, there are a large number of centralized installations [7]. In Africa and Latin American countries, the related facilities in the literature are scarce, there are several technologies to improve biogas, but they have disadvantages such as high energy consumption or high initial investment costs [8]. For this reason, the scientific community has developed studies to improve the efficiency of the processes. In 2020, Roubik and Mazancová presented a project for the application of biogas from cattle manure to meet the energy needs of households located in rural areas [9]. A year later, Arutyunov et al. presented a proposal for the application of low-scale biogas plants for biogas production, as well as liquid biofuels for application in agricultural areas [10]. In turn, Lohani et al. presented a case study in Nepal where biogas is used for cooking, benefiting several poor communities [11]. From another angle, O'Connor et al. presented a review of the current status of small-scale biogas production systems on farms in Europe, discussing the current status and production capacity on the continent [12].

Currently, alternative processes are being developed in order to improve the potential of biogas in terms of CH_4 percentages, which include the use of excess CO_2 in the presence of H_2O for absorption system [13]. Angelidaki et al, developed a study where they transform H_2 into CH_4, considered as appropriate and as an alternative for utilization as natural gas [14]. Other methods are under development, such as biogas recirculation inside the biodigester allowing in situ biogas upgrading, being more cost-effective and simpler than commercial upgrading processes [15].

The objective of this study is to identify the technologies and processes for biogas production at low and micro scale, taking into account sustainability characteristics that allow evaluating, according to the context, their possible implementation. The information analysis will be developed under the application of the formal concept analysis method [16] [17][18] and the use of the Concept Explorer [19] and VOSviewer [20], software, relating a series of objects and attributes, which will facilitate the processing of the information. With this approach in mind, the document is organized as follows. Section 2 presents the methodology, defining the attributes and assigning each biogas production technology an object for the development of the formal concept analysis. Section 3 covers the literature review, focusing on the level of compliance with each of the attributes. Section 4 presents the critical review using the Concept Explorer tool and the formal concept analysis method, evaluating small and micro-scale technologies. Finally, Section 5 presents the most relevant conclusions of the work.

2. Methodology

With the intention of evaluating technologies that produce biogas anaerobically at low and micro scale, Figure 1 presents the methodology for the development of this study, initially a search for information is performed using scientific databases such as Dimensions, Lens, Scopus and Web Of Science, among others, with the following search string; "biogas* production*" OR "Anaerobic* Digestion*" OR "Low* Scale" AND "Review*", where manuscripts were extracted and subsequently analyzed by VOSviewer software allowing to establish a bibliographic coupling network that correspond by means of keywords and authors with higher citations (Figure 1). The criteria for selecting the information were defined by the year of publication and the percentile of the journals in which the manuscript was published; 56 articles and 18 book chapters were selected.

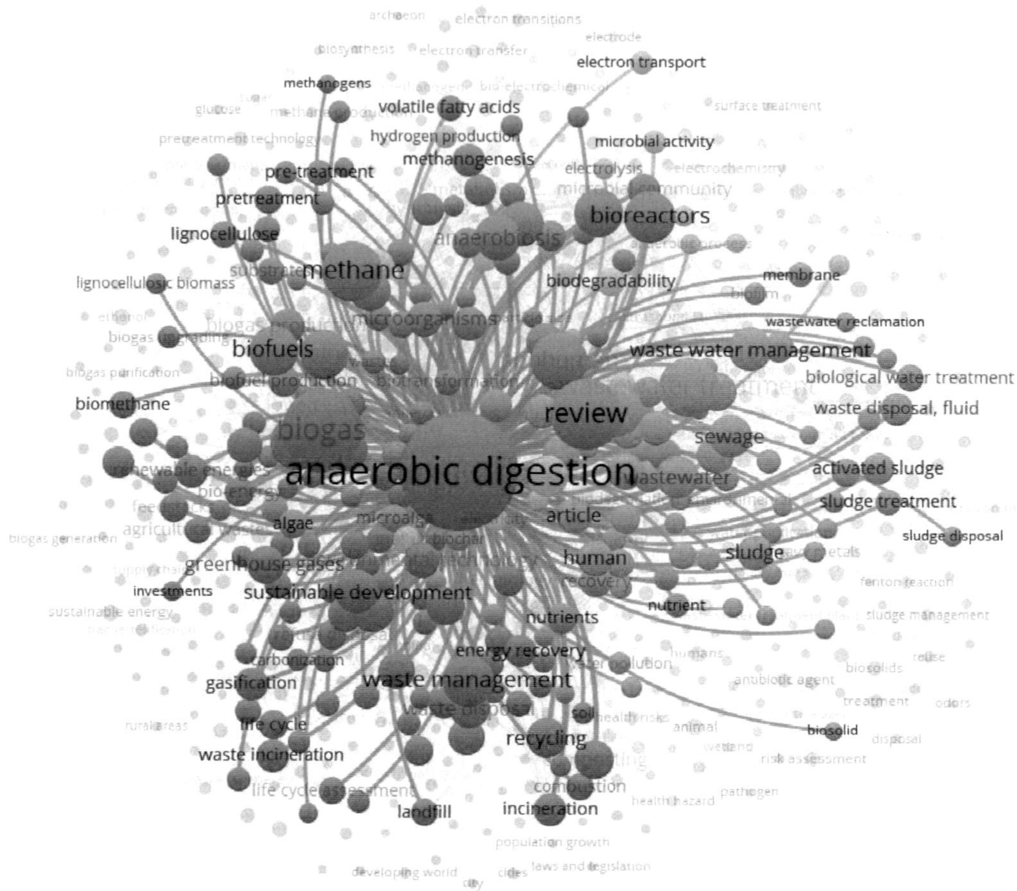

Figure 1. Bibliographic linkage network Keywords / Authors

In the application of the FCA methodology, objects and attributes must be defined [21], for this study the objects were defined in micro-scale and low-scale biogas production technology and the attributes that each biogas production technology must meet are:

A1.Life cycle: will meet the object with a more favorable life cycle assessment (LCA).

A2.Implementation area: meets the object with the lowest environmental impact on land use for operation, with high performance levels.

A3. Application of resources or solid wastes (Waste) for its production.

A4. Efficiency of the system: will meet the object with the highest level of efficiency.

A5. Cost: the object with the lowest initial investment cost will meet the criterion.

A6. Technology complexity: lower maintenance and operation requirements.

A7. Maturity: the object with the highest level of maturity will meet the criterion.

Taking into account the objects of this study, a literature review is carried out according to the compliance to a greater or lesser extent of each attribute. This will result in a matrix that will allow a qualitative analysis. The data analysis will be performed through the open access software "Concept Explorer"[19], evaluating all existing relationships, grouping technologies and attributes.

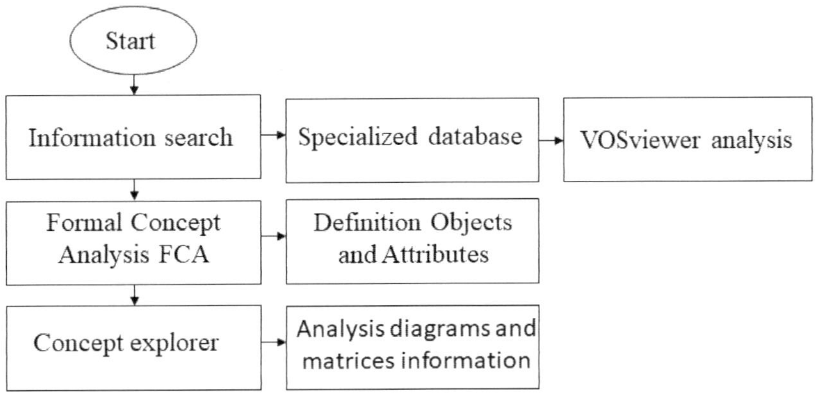

Figure 2. Methodology used in the development of the study.

3. Low- and micro-scale biogas production

This section presents theoretical concepts to classify the types of biodigesters, different biogas production technologies and the respective classification according to energy production.

3.1 Anaerobic digestion

It is called biomethanization or biogas production. Generally, an anaerobic process is a highly complex microbial process that takes place in the absence of oxygen, where organic matter is decomposed by the action of microorganisms, generating a mixture of gases (biogas) consisting mainly of methane and carbon dioxide[22].

The organic matter is decomposed by different groups of microorganisms into simple compounds, which become volatile fatty acids. These acids are consumed by methanogenic microorganisms that result in the production of methane and carbon dioxide. This whole process is carried out in a biodigester or reactor [23]. Table 1 shows four different stages in the anaerobic processes

Table 1. Stages of the anaerobic digestion process.

Stage	Process	Characteristic
1	**Hydrolisis**	Bacteria that break the chain into smaller constituent parts. Complex organic molecules are broken down into simple sugars, amino acids and fatty acids.
2	**Acidogenesis**	Bacteria produce acid to break down simple sugars, amino acids and fatty acids into carbon dioxide.
3	**Acetogenesis**	This process produces mainly acetic acid
4	**Methanogenesis**	Methane is produced. More carbon dioxide and water are produced. Methanogenesis is sensitive to pH between 6.5 and 8.

Source: Author. Adapted from [22]

3.2. Biogas production technologies

There are several technologies used in the biogas system[24] [25], Additionally, for decentralized systems Rasimphi et al. classifies plants into fixed dome, floating drum, balloon, horizontal, earthen pit and ferrocement [7] [26]. Henceforth, the term decentralized system refers to small-scale biogas plants, linking small- and micro-scale systems into one category [6]. The designs are generally supported by geographical location, by the availability of materials in the local area for technological implementation [26][25]28].

Table 2. Type of reactors for biogas production

Reactor type	Reactor Characteristic	Ref
Continuous flow stirred tank (CSTR)	It could agitate biogas generation and accelerate digestion progress.	[28]
Anaerobic Plug Flow Reactor (APFR)	Compared with single-stage STRs, APFRs can make the flow rate more stable	[29]
Anaerobic contact (ACR)	Can mix the tank with sludge recycle	[27]
Batch	Widely used in small-scale farms.	[27]
Hybrid bioreactor	Weaknesses of batch reactors are channeling and clogging.	[27]

Source: Author

3.3. Plant size classification

Based on the size of the combined heat and power (CHP) system biogas plants can be classified into micro, small, medium and large scale. Micro-scale has a CHP of less than 15kWe. Small scale is between 15kWe and 99kWe. Medium scale is between 100kWe and 299kWe and large scale exceeds 300kWe. The difference of the scale is also different in the design of the systems. The first is the passive system which is defined as a plant where the biogas recovery unit is in the existing biomass. The temperature is below 20 ◦C. The second is the low rate system, where the digesters need more time to digest. The third system is the high rate system which requires digesting liquid in a short period and keeping the solid in a longer period, but still shorter than the low rate system[12]. Table 3 presents a classification of biogas plants according to their energy capacity.

Table 3. Sizing classification of biogas plants by anaerobic digestion

Size	Capacity	Description	Applications
Micro Scale	Cogeneration electric power between 5-15 kW	It is limited to the treatment of small quantities of organic waste.	-Domestic use
Low scale	Cogeneration electric power between 15 - 99 kW	Uses biomass generally available on farms for heat and power production.	-Local heating
Medium scale	Cogeneration electrical capacity between 100-299 kW	Linking plants between small and large scale.	-Agricultural farms
Large scale	Cogeneration Electrical power> 300 kW	Generally, have a feedstock consumption exceeding 5000 tons per year.	-Medium and large communities

Source: Author. Adapted from [32][12]

Small-scale biogas technologies provide a number of benefits to communities in isolated areas [33]. These benefits range from socioeconomic to environmental areas, and include improved health and air quality in the environment where the system is implemented, as well as improved fertilizer quality [34]. Additionally, it generates employment by using local labor for its manufacture and operation over time, as well as diversification in the supply of electricity for the area [35]. Micro and small-scale biogas technology is crucial for sustainable development through circular economy schemes since they are generally fed with animal manure, energy crops and solid wastes to address the lack of energy in impoverished rural communities, small-scale biogas is a viable alternative to remedy this problem to some extent [36] [37].

It is important to mention that farmers must take into account the leaching of biogas production into the aquatic environment in order to generate systems that are sufficiently sustainable in environmental terms, i.e., the use of soils is an important item for the proper management of the by-products of the process [38] [39]. Namely, the size of the small-scale system can generate greater or lesser impact on soil use, however, the issue should focus on the proper management of leachate regardless of whether the plant has low or micro-scale characteristics, i.e., the impact can be the same regardless of the size of the system [39] [7].

The economics of small-scale systems comprise lower investment costs with respect to centralized systems [40], however, the size of the plant directly influences the initial cost of investment, maintenance and operation [41], that is to say, the micro-scale system will have a lower investment value than low-scale plants [42]. This economic factor is evaluated taking into account that the use of raw material for these two types of systems is free, and comes, for example, from animal manure, water, treated sludge, solid waste, agricultural waste, among others [43].

In terms of efficiency, both small and micro-scale plants vary in efficiency based on the use of resources in the area [44]; the type of feedstock used in the process [45]. The feedstock varies according to the geographical area of implementation and the available resources [46]. However, low-scale systems, having a higher production capacity, present higher efficiencies [47]. In terms of life cycle and maturity, low-scale systems are generally manufactured with higher strength materials than micro-scale plants and have a larger number of applications worldwide [48] [49] [50].

Finally, Table 4 presents the matrix of attributes for the two types of biogas plants; low and micro scale. The attributes that meet each attribute are identified with an X in the matrix.

Table 4. Selection matrix of attributes with respect to objects

Ob \ At	A1	A2	A3	A4	A5	A6	A7
Low Scale	X	X	X	X			X
Micro Scale		X	X		X	X	

Source: Author

4. Results

Figure 2 represents the graphical diagram of the formal concept analysis performed in the Concept Explorer tool to evaluate low and micro-scale biogas plants, based on the information provided in the attribute's matrix in Table 6. This diagram relates the small-scale plants; classified in low and micro scale with those 7 attributes proposed to identify their advantages and disadvantages as an alternative technology.

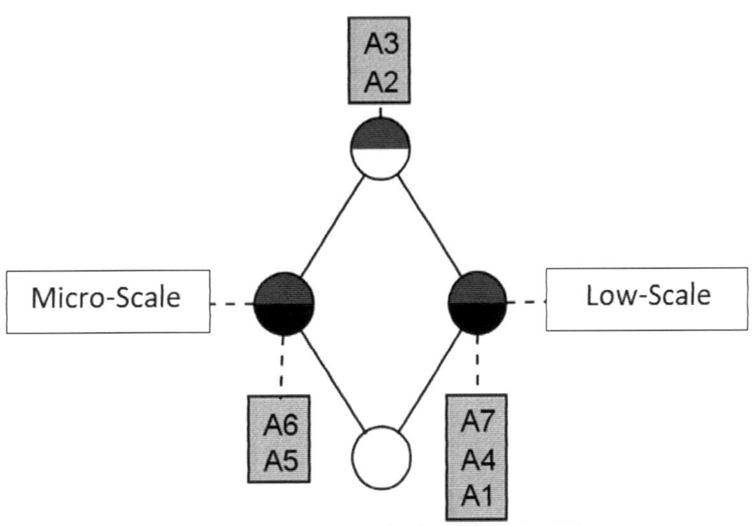

Figure 3. Diagram of information analysis FCA methodology

The different nodes shown in Figure 2 are identified as follows:

i. Blue-white: contains only attributes
ii. Blue-black: they are containers of micro and low scale technologies.
iii. White: connector node

Considering the above, the analysis of the diagram is simple. The low and micro-scale biogas plants analyzed are related to all the attributes that connect them from top to bottom, up to the node of origin. Specifically, attribute A2 "Area of implementation" and A3 "Application of resources or solid waste for production" are shared by the two types of biogas plants, i.e., they comply with this attribute. Additionally, low-scale biogas plants comply with three attributes individually; A1 "Life cycle", A4 "System efficiency" and A7 "Maturity". On the other hand, micro-scale biogas plants do not comply with the above mentioned tributes, however, they comply with two attributes individually; A5 "Cost" and A6 "Complexity of the technology". From this analysis, it is evident that both types of biogas plants have favorable characteristics to provide energy solutions at a decentralized level for isolated or rural areas. As an alternative to the theoretical analysis approach, we can state that low scale systems are more suitable for small scale biogas plant applications, however, micro scale biogas plants can overcome the attributes that it does not meet, through the development of new systems, the linkages of new design materials and biomass application in a suitable way in the medium and long term.

5. Conclusions

The information was analyzed under the FCA methodology, evaluating characteristics such as costs, efficiency, land use, life cycle, maturity, among others, contributing to overcome current problems for a specific problem. The most relevant conclusions of the work are presented below:

i. Most of the literary information does not present information on the classification of biogas plants based on their size, so there is a need for a classification at a global level.
ii. There is a high interest by different authors in the characterization of solid wastes or biomass used in the process of biogas production at low and micro scale. Animal manure and solid wastes (garbage) are used by different authors and reach acceptable biogas production levels for each specific analysis presented in their publications.
iii. Low-scale biogas plants comply with 5 of the 7 proposed attributes, showing greater applications than micro-scale systems. However, both systems are highly applied in small North America, Europe, Asia and especially in Canada, for electric energy supply.
iv. In general, small-scale biogas plants are an alternative for the production of energy and the use of environmentally polluting matter, i.e., they solve waste treatment problems and produce biogas that can be used as a source of energy. However, the treatment of the leachate produced in the process must be considered for the system to be environmentally sustainable.

Translated with www.DeepL.com/Translator (free version)the technological development of low-scale and micro-scale commercial systems, which allows generating a system with rural application at low cost and with acceptable efficiency, generating a socio-economic impact in impoverished communities that lack energy resources; heating and electricity. The current centralized systems are highly expensive, so there is a challenge that involves reducing costs in design, construction and implementation, as well as maintenance and operation over time.

References

[1] Q. Yu, R. Liu, K. Li, and R. Ma, "A review of crop straw pretreatment methods for biogas production by anaerobic digestion in China," *Renew. Sustain. Energy Rev.*, vol. 107, pp. 51–58, Jun. 2019.

[2] K. Vohra, A. Vodonos, J. Schwartz, E. A. Marais, M. P. Sulprizio, and L. J. Mickley, "Global mortality from outdoor fine particle pollution generated by fossil fuel combustion: Results from GEOS-Chem," *Environ. Res.*, vol. 195, p. 110754, Apr. 2021.

[3] S. Kaza, L. Yao, P. Bhada-Tata, and F. Van Woerden, "What a waste 2.0. A Global Snapshot of Solid Waste Management to 2050," Washington, DC: World Bank, 2018.
[4] W. Ferdous, A. Manalo, R. Siddique, P. Mendis, Y. Zhuge, H. S. Wong, W. Lokuge, T. Aravinthan, and P. Schubel, "Recycling of landfill wastes (tyres, plastics and glass) in construction – A review on global waste generation, performance, application and future opportunities," *Resour. Conserv. Recycl.*, vol. 173, p. 105745, Oct. 2021.
[5] T. E. Rasimphi, D. Tinarwo, S. Ravhengani, C. Sambo, and P. Mhlanga, "Chapter 26 - Biogas technology implementation in rural areas: a case study of Vhembe District in Limpopo Province, South Africa," in *Handbook of Biofuels*, S. Sahay, Ed. Academic Press, 2022, pp. 501–509.
[6] K. G. Wilkinson, "A comparison of the drivers influencing adoption of on-farm anaerobic digestion in Germany and Australia," *Biomass and Bioenergy*, vol. 35, no. 5, pp. 1613–1622, May 2011.
[7] T. E. Rasimphi, D. Tinarwo, C. Sambo, M. A. Mutheiwana, and P. Mhlanga, "Chapter 24 - Decentralized biogas plants: status, prospects, and challenges," in *Handbook of Biofuels*, S. Sahay, Ed. Academic Press, 2022, pp. 473–484.
[8] M. Hiloidhari and S. Kumari, "Chapter 15 - Biogas upgrading and life cycle assessment of different biogas upgrading technologies," in *Emerging Technologies and Biological Systems for Biogas Upgrading*, N. Aryal, L. D. Mørck Ottosen, M. V. Wegener Kofoed, and D. Pant, Eds. Academic Press, 2021, pp. 413–445.
[9] H. Roubík and J. Mazancová, "Suitability of small-scale biogas systems based on livestock manure for the rural areas of Sumatra," *Environ. Dev.*, vol. 33, p. 100505, 2020.
[10] V. Arutyunov, A. Nikitin, L. Strekova, V. Savchenko, and I. Sedov, "Utilization of renewable sources of biogas for small-scale production of liquid fuels," *Catal. Today*, vol. 379, pp. 23–27, Nov. 2021.
[11] S. P. Lohani, B. Dhungana, H. Horn, and D. Khatiwada, "Small-scale biogas technology and clean cooking fuel: Assessing the potential and links with SDGs in low-income countries – A case study of Nepal," *Sustain. Energy Technol. Assessments*, vol. 46, p. 101301, Aug. 2021.
[12] S. O'Connor, E. Ehimen, S. C. Pillai, A. Black, D. Tormey, and J. Bartlett, "Biogas production from small-scale anaerobic digestion plants on European farms," *Renew. Sustain. Energy Rev.*, vol. 139, p. 110580, Apr. 2021.
[13] J. S. Sravan, A. Tharak, and S. V. Mohan, "Chapter 1 - Status of biogas production and biogas upgrading: A global scenario," in *Emerging Technologies and Biological Systems for Biogas Upgrading*, N. Aryal, L. D. Mørck Ottosen, M. V. Wegener Kofoed, and D. Pant, Eds. Academic Press, 2021, pp. 3–26.
[14] I. Angelidaki, L. Xie, G. Luo, Y. Zhang, H. Oechsner, A. Lemmer, R. Munoz, and P. G. Kougias, "Chapter 33 - Biogas Upgrading: Current and Emerging Technologies," in *Biofuels: Alternative Feedstocks and Conversion Processes for the Production of Liquid and Gaseous Biofuels (Second Edition)*, A. Pandey, C. Larroche, C.-G. Dussap, E. Gnansounou, S. K. Khanal, and S. Ricke, Eds. Academic Press, 2019, pp. 817–843.
[15] M. Miltner, A. Makaruk, and M. Harasek, "Review on available biogas upgrading technologies and innovations towards advanced solutions," *J. Clean. Prod.*, vol. 161, pp. 1329–1337, Sep. 2017.
[16] B. Ganter and R. Wille, *Análisis de concepto formal, Texto original: Fundamentos matemáticos*. Berlín Heidelberg: Springer-Verlag, 1999.
[17] F. Sancho Caparrini, *Análisis Formal de Conceptos*. 2017.
[18] B. E. Tarazona-Romero, A. Campos-Celador, Y. A. Muñoz-Maldonado, J. G. Ascanio-Villabona, M. A. Duran-Sarmiento, and A. D. Rincón-Quintero, "Development of a Fresnel Artisanal System for the Production of Hot Water or Steam," in *Recent Advances in Electrical Engineering, Electronics and Energy*, 2021, pp. 196–209.

[19] S. Yevtushenko, J. Tane, T. B. Kaiser, S. Objedkov, J. Hereth Correia, and H. Reppe, *Concept Explorer*. 2021.
[20] N. Van Eck and L. Waltman, "VOSviewer Visualizing scientific landscapes." 2022.
[21] B. E. Tarazona-Romero, A. Campos-Celador, and Y. A. Maldonado-Muñoz, "Can solar desalination be small and beautiful? A critical review of existing technology under the appropriate technology paradigm," *Energy Res. Soc. Sci.*, vol. 88, p. 102510, 2022.
[22] T. Kunatsa and X. Xia, "A review on anaerobic digestion with focus on the role of biomass co-digestion, modelling and optimisation on biogas production and enhancement," *Bioresour. Technol.*, p. 126311, Nov. 2021.
[23] R. A. Parra Huertas, "Digestión anaeróbica: mecanismos biotecnológicos en el tratamiento de aguas residuales y su aplicación en la industria alimentaria," vol. 10, no. 2, 2015.
[24] A. Omer, "Biogas technology for sustainable energy generation: development and perspectives," *MOJ Appl. Bionics Biomech.*, vol. Volume 1, no. Issue 4, Nov. 2017.
[25] A. I. Adnan, M. Y. Ong, S. Nomanbhay, K. W. Chew, and P. L. Show, "Technologies for Biogas Upgrading to Biomethane: A Review," *Bioengineering*, vol. 6, no. 4, p. 92, Dec. 2019.
[26] A. Ahmad, D. Andrio, T. Y. Putra, and U. Seprizal, "Determination of mass transfer coefficient from biogas purification process by mono ethanol amine (MEA)," *Mater. Today Proc.*, Mar. 2022.
[27] Q. Xu, Y. Tian, S. Wang, and J. H. Ko, "A comparative study of leachate quality and biogas generation in simulated anaerobic and hybrid bioreactors," *Waste Manag.*, vol. 41, pp. 94–100, Jul. 2015.
[28] M. Zhang, J. Gao, Y. Fan, Q. Liu, C. Zhu, L. Ge, C. He, and J. Wu, "Comparisons of nitrite accumulation, microbial behavior and nitrification kinetic in continuous stirred tank (ST) and plug flow (PF) moving bed biofilm reactors," *Chemosphere*, vol. 278, p. 130410, Sep. 2021.
[29] D. B. Panaro, M. R. Mattei, G. Esposito, J. P. Steyer, F. Capone, and L. Frunzo, "A modelling and simulation study of anaerobic digestion in plug-flow reactors," *Commun. Nonlinear Sci. Numer. Simul.*, vol. 105, p. 106062, Feb. 2022.
[30] E. Şentürk, M. İnce, and G. O. Engin, "The effect of transient loading on the performance of a mesophilic anaerobic contact reactor at constant feed strength," *J. Biotechnol.*, vol. 164, no. 2, pp. 232–237, Mar. 2013.
[31] E. L. Odekanle, S. O. Dahunsi, and S. Zahedi, "Anaerobic treatment of abattoir waste: Biogas production and correlation parameter in a batch reactor system," *J. Water Process Eng.*, vol. 37, p. 101337, Oct. 2020.
[32] M. Walker, H. Theaker, R. Yaman, D. Poggio, W. Nimmo, A. Bywater, G. Blanch, and M. Pourkashanian, "Assessment of micro-scale anaerobic digestion for management of urban organic waste: A case study in London, UK," *Waste Manag.*, vol. 61, pp. 258–268, Mar. 2017.
[33] S. Chelliapan, N. Arumugam, M. F. Md. Din, H. Kamyab, and S. S. Ebrahimi, "Chapter 11 - Anaerobic treatment of municipal solid waste landfill leachate," in *Bioreactors*, L. Singh, A. Yousuf, and D. M. Mahapatra, Eds. Elsevier, 2020, pp. 175–193.
[34] O. V. Okoro, Z. Sun, and J. Birch, "10 - Thermal depolymerization of biogas digestate as a viable digestate processing and resource recovery strategy," in *Advances in Eco-Fuels for a Sustainable Environment*, K. Azad, Ed. Woodhead Publishing, 2019, pp. 277–308.
[35] P. Ghosh, G. Shah, S. Sahota, L. Singh, and V. K. Vijay, "Chapter 7 - Biogas production from waste: technical overview, progress, and challenges," in *Bioreactors*, L. Singh, A. Yousuf, and D. M. Mahapatra, Eds. Elsevier, 2020, pp. 89–104.
[36] F. Xu and Y. Li, "Biomass Digestion," in *Encyclopedia of Sustainable Technologies*, M. A. Abraham, Ed. Oxford: Elsevier, 2017, pp. 197–204.
[37] A. Dahiya, "Chapter 18 - Gaseous fuels and bioelectricity service learning projects and case studies," in *Bioenergy (Second Edition)*, A. Dahiya, Ed. Academic Press, 2020, pp. 361–380.
[38] F. Xu, A. Khalaf, J. Sheets, X. Ge, H. Keener, and Y. Li, "Chapter Three - Phosphorus Removal and Recovery From Anaerobic Digestion Residues," in *Advances in Bioenergy*, vol.

3, Y. Li and X. Ge, Eds. Elsevier, 2018, pp. 77–136.
[39] Z. Youcai and W. Ran, “Chapter 5 - Technical development and a pilot trial for anaerobic digestion of water hyacinth,” in *Biomethane Production from Vegetable and Water Hyacinth Waste*, Z. Youcai and W. Ran, Eds. Elsevier, 2021, pp. 265–335.
[40] M. Fedeli and F. Manenti, “Assessing Process Effectiveness with Specific Environmental and Economic Impact of Heat, Power & Chemicals (HPC) Option as Future Perspective in Biogas,” *Clean. Chem. Eng.*, p. 100016, Apr. 2022.
[41] C.-C. Cormos, A.-M. Cormos, L. Petrescu, and S. Dragan, “Techno-economic assessment of decarbonized biogas catalytic reforming for flexible hydrogen and power production,” *Appl. Therm. Eng.*, vol. 207, p. 118218, May 2022.
[42] A. W. Yalew, “Economic contributions and synergies of biogas with the SDGs in Ethiopia,” *Energy Nexus*, vol. 3, p. 100017, Dec. 2021.
[43] S. V. Dewalkar and S. S. Shastri, “Life cycle environmental and economic assessment of human night soil based Malaprabha biogas plant,” *Mater. Today Proc.*, Dec. 2021.
[44] Z. K. Liew, Y. J. Chan, Z. T. Ho, Y. H. Yip, M. C. Teng, A. I. T. Ameer Abbas bin, S. Chong, P. L. Show, and C. L. Chew, “Biogas production enhancement by co-digestion of empty fruit bunch (EFB) with palm oil mill effluent (POME): Performance and kinetic evaluation,” *Renew. Energy*, vol. 179, pp. 766–777, Dec. 2021.
[45] S. P. Ramachandran, A. Shrivastava, A. Sharma, M. Feroskhan, R. Manoj Kuma, and S. M. Dasharath, “Performance, emission and combustion characteristics of a biogas–diesel dual fuel engine using Taguchi method,” *Mater. Today Proc.*, vol. 54, pp. 548–556, Jan. 2022.
[46] D. Papurello, S. Silvestri, and S. Modena, “Biogas trace compounds impact on high-temperature fuel cells short stack performance,” *Int. J. Hydrogen Energy*, vol. 46, no. 12, pp. 8792–8801, Feb. 2021.
[47] M. Ishaq and H. Ishaq, “Performance assessment of biogas-fed solid oxide fuel cell system for municipal solid waste treatment,” *J. Clean. Prod.*, p. 131702, Apr. 2022.
[48] W. Zhao, H. Yang, S. He, Q. Zhao, and L. Wei, “A review of biochar in anaerobic digestion to improve biogas production: Performances, mechanisms and economic assessments,” *Bioresour. Technol.*, vol. 341, p. 125797, Dec. 2021.
[49] M. B. Jensen, S. Poulsen, B. Jensen, A. Feilberg, and M. V. W. Kofoed, “Selecting carrier material for efficient biomethanation of industrial biogas-CO2 in a trickle-bed reactor,” *J. CO2 Util.*, vol. 51, p. 101611, Sep. 2021.
[50] L. Lehmann and E. Bloem, “Antibiotic residues in substrates and output materials from biogas plants – Implications for agriculture,” *Chemosphere*, vol. 278, p. 130425, Sep. 2021.

IOP Conf. Series: Materials Science and Engineering 1299 (2024) 012009 doi:10.1088/1757-899X/1299/1/012009

Concentrating Solar Power Systems: a review of their applicability in Colombia

B. E. Tarazona-Romero[1 3], K. T. Jaimes-Quintero[1], M. D. Sarmiento-Duran[2], J. G. Ascanio-Villabona[1 2] and E. F. Amaya-Arias

[1] Automation and Control Energy Systems Research Group (GISEAC), Faculty of Natural Sciences and Engineering, Electromechanical engineering, Unidades Tecnologicas de Santander (UTS), Student Street No 9-82, 680005, Bucaramanga, Colombia.
[2] Energy in Building research group (ENEDI), Doctoral Program in Energy Efficiency and Sustainability in Engineering and Architecture, Department of Machines and Thermal Engines, University of the Basque Country (UPV / EHU), Engineer Torres Quevedo Square, 1, 48013, Bilbao, BI, Spain.

E-mail corresponding autor:btarazona@correo.uts.edu.co

Abstract. This paper evaluates the applicability of centralized concentrating solar power systems in Colombia. For this purpose, these systems have been classified into four subsystems: parabolic trough, linear Fresnel, parabolic dish and tower. Each is explored for its suitability by meeting a set of six attributes. A bibliometric analysis process was developed using the Scopus database and the VOSviewer software to present the potential of each subsystem evaluated through an evaluation matrix. The analysis of the subsystems and attributes was carried out through the formal concept analysis methodology 8FCA). To facilitate data processing, the open access software concept Explorer is implemented. The analysis shows that parabolic trough technology is more widely used worldwide than the other three technologies evaluated, and its maturity makes it possible to plan centralized and decentralized projects in developing countries. The bibliometric analysis integrated with the applied FCA methodology has proven to be useful for the evaluation of technological typologies and serves as an alternative to develop theoretical studies that group and link different options as a model to evaluate a deterministic set of attributes.

Keywords: Bibliometric analysis, Formal Concept Analysis, Solar Concentrators

1. Introduction

Climate change is perhaps one of the greatest challenges facing the human species today and may be the result of natural causes or human influence. The latter produces the climatic alterations with the highest levels of incidence due to the use of fossil fuels for decades, affecting diverse weather patterns that directly incur in problems such as global warming. However, human beings must maintain a quality of life and economic development; therefore, the use of fossil fuels has been helpful for human survival and development [1].

The use of fossil fuels has increased in recent decades with industrial development and population growth. It is estimated that in one-year human beings consume what nature has processed for a million years and currently, more is consumed than nature is able to process, projecting an irreversible shortage in the future. Additionally, its use generates problems such as: alteration of the greenhouse effect, acid rains, social problems worldwide, among others. [2].

According to the International Energy Agency, in the year 2021 the emissions of Carbon Oxide (CO_2) related to energy will grow by 5%, being CO_2 one of the gases with the highest generation in the world due to the burning of fossil fuels and, at the same time, the one that has the greatest impact on global warming [3]. In order to stabilize gas concentrations in the world, the United Nations Conference on Climate Change was held at the United Nations Headquarters in New York, USA. (CP26), was held to review the actions of the countries worldwide that adhered to the Paris Agreement (COP21) and to ratify the commitment of all to reduce the consumption of fossil fuels and the use of renewable energies as an alternative to mitigate climate change [2].

Consequently, the use of renewable energies has increased in the last decade and technologies that take advantage of sources such as: sun, wind, biomass, water, heat, among others, have been implemented in some sectors worldwide in a centralized and decentralized manner for energy production, highlighting the use of solar radiation as one of the main sources of electricity and heat generation. However, the process is increasing in developed countries, with the opposite scenario in developing countries [4].

Currently, the largest power generation plants using renewable resources use photovoltaic panel technology [5]. However, there are technologies such as concentrating solar power systems that also take advantage of solar radiation and are a viable alternative for centralized energy production in the world (largest number of plants in the U.S. and Spain), projected to be able to supply 6% of the world's demand in 2030 and 12% in 2050 [6].

For the theoretical development of this document, the solar concentration technologies will be evaluated, applying a bibliometric analysis by means of VOSviewer software, the information obtained will be evaluated by means of the method of formal analysis of concepts [7] [8] and the use of the software concept Explorer [9], relating attributes and objects, simplifying information processing. With this in mind, the objectives of this work are:

i. Define the attributes that satisfy each of the concentrating solar technologies.
ii. Develop a review of the academic literature through databases and specialized software to evaluate the objects under the established attributes.
iii. Discuss the relationship between the technology and the attributes established through the use of formal Concept Analysis.

For this approach, the document is organized as follows. Section 2 presents the methodology, defining the bibliometric search string, as well as the attributes and assigning to each technology an object for the development of the formal concept analysis. Section 3 covers the literature review on solar concentrators, focusing on the level of compliance with each of the attributes. Section 4 presents the critical review using the Concept Explorer tool and the formal concept analysis method. Finally, section 5 presents the most relevant conclusions of the study.

2. Methodology

The formal analysis of concepts is an objective analysis based on a series of attributes and objects, which allows us to relate them and determine which objects comply with a greater number of attributes, showing the attributes they share among them [8]. Consequently, the existing types of CSP solar concentrators will be related as objects and it will be analyzed whether they meet the following attributes in order to comply with the objectives outlined. It is important to note that, to a certain extent, the author of this paper has taken the following attributes as guidelines to identify the technology with the greatest potential worldwide that adapts to Colombia's needs.

Initially, a specialized bibliometric analysis was carried out to identify the articles with the highest incidence in the subject of study in order to feed the FCA methodology. The database used was Scopus and the program used to manage the information was the VOSviewer software, as shown in Figure 1.

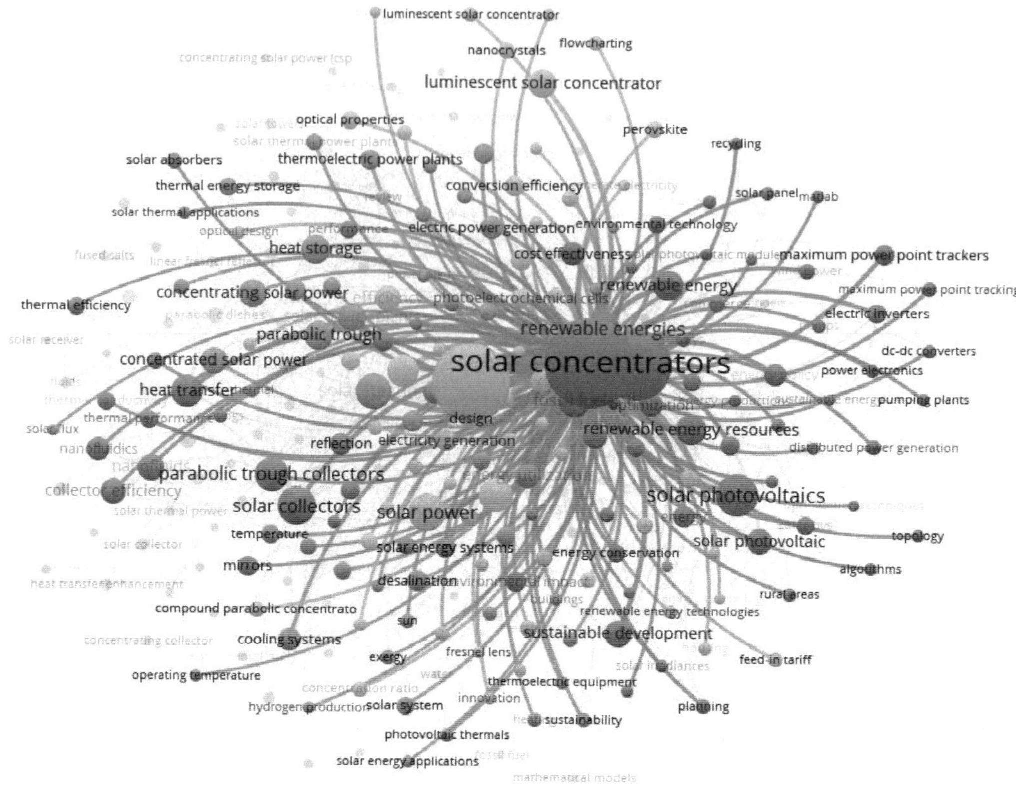

Figure 1. Bibliometric Analysis

For its part, the formal analysis of concepts is based on the fulfillment, to a greater or lesser extent, of a series of attributes with respect to a series of objects [13]. The following are the attributes that each carbon sequestration technology system must meet:

A1. Cost: will meet the criterion of the two technologies with the lowest initial investment cost.

A2. Technology complexity: the two technologies with the lowest maintenance and operation requirements will meet the criterion.

A3. Maturity: will meet the highest maturity level of the four technologies.

A4. Life cycle: will the technology comply with a more favorable life cycle assessment (LCA)

A5. Implementation area: meet the two technologies with the lowest land use for operation, with high yield levels

A6. Weather Conditions

A7. System efficiency: the two most efficient technologies will meet the highest level of efficiency; the concentration ratio will be measured.

Considering the objects and attributes, each object is evaluated by each of the attributes listed above. This will result in a matrix that will feed the concept Explorer tool, allowing a qualitative evaluation of the information [9].

3. Concentrating Solar Power Technologies

3.1. Global Warming in the world

Today, the planet is highly dependent on solutions to existing environmental problems that have become more acute in recent decades [10]. Soil contamination, air pollution, scarcity of drinking water, population increase, decrease of fertile land for cultivation, deforestation, cattle raising, among others, are direct causes of climate change and global warming [11].

The need for energy access throughout the world to maintain quality of life and technological development requires a supply of electrical and thermal energy. To cover this energy demand, an accelerated consumption of fossil fuels has been generated, generating as a negative effect the production of gases that affect the greenhouse effect, aggravating and precipitating global warming [1], environmental challenge facing the planet today.

The condition of the environment is one of the major concerns of governments around the world, due to the high levels of pollution that have an immense impact on people's quality of life [12]. The industrial revolution was the turning point at which greenhouse gas (GHG) emissions into the atmosphere began to skyrocket. Figure 2 shows the trend of CO_2 accumulation from 1958 to 2020 and shows a marked increase in recent decades, with CO_2 being one of the gases produced by the burning of fossil fuels and perhaps the gas with the greatest impact on global warming [2].

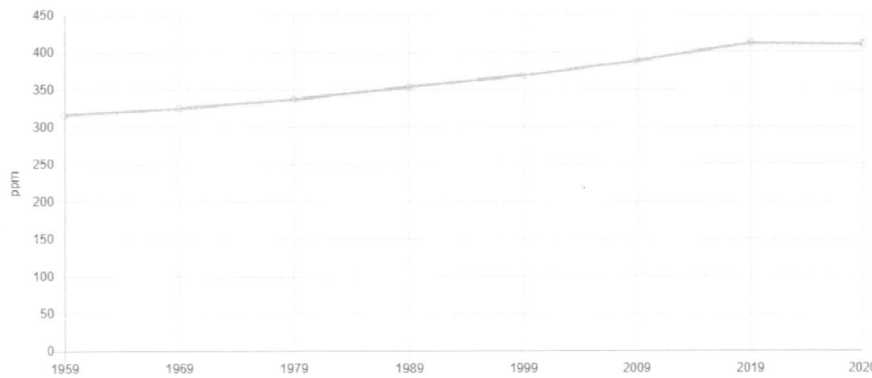

Figure 2. Atmospheric CO_2 Accumulation Trend 1959-2020 [2]

GHGs are produced naturally and are essential to maintain the planet earth in ideal conditions for humanity to survive, however, the excessive production of gases, deforestation and large-scale agriculture have directly intervened in this natural process, causing environmental phenomena that directly affect human life. he most abundant GHG is carbon dioxide (CO_2), which results from the burning of fossil fuels, used for example worldwide for the mobility of transport vehicles [13].

Based on the report presented by Foster and Elzinga on the UN website it is highlighted that: "Fossil fuels currently comprise approximately 80% of current global primary energy demand and the energy system is the source of approximately two-thirds of global CO_2 emissions" [1].

Foster y Elzinga, emissions do not include emissions of methane and other short-lived climate pollutants, so energy production and use may be sources with higher proportions of emissions. In turn, the UN Intergovernmental Panel on Climate Change (IPPC) was created by the World Meteorological Organization (WMO) and UN Environment to provide an objective view of scientifically reliable information on the status and future of global warming. The IPPC presented a report in the year 2021, entitled "Climate Change 2021: Physical Basis", where they state: "Global warming is Widespread, rapidly advancing and intensifying. They also underline the urgency of substantial, rapid and sustained reductions in greenhouse gas emissions. In addition, they point out that in all regions of the world the climate system is changing and that phenomena such as the continuous rise in sea level will not be reversed for several centuries" [14].

The report presented by the IPPC makes it clear that human action in current environmental problems is indisputable, and highlights that humanity must consider substantial and sustained measures to immediately reduce emissions of carbon dioxide (CO_2) and other greenhouse gases to curb the growing pollution of the atmosphere, generating immediate health benefits, stabilizing the global average temperature within two to three decades [14].

The IPPC report, endorsed by 195 governments and prepared by Working Group I, is the first installment of the Sixth Assessment Report "AR6 Climate Change 2021: The Physical Science Basis" and will be completed by the end of 2022. For the first time, it is expected to complete a detailed analysis of climate change at the regional level, allowing to show the current state and help interpret what is supposed to be the appropriate path for human action to generate a positive impact on environmental problems in society and ecosystems.

If current trends of accelerated fossil fuel use continue due to energy demand that is projected to double by 2050, gas emissions will far exceed the levels proposed to curb global warming and could instead increase global temperatures by 2°C [1]. However, the reduction of emissions should not exclude the use of fossil fuels, but requires a significant change of direction, i.e. the use of renewable energy technologies, improved energy efficiency in industrial facilities and processes worldwide, as well as carbon capture and storage [15].

This is why the assumption of using only fossil fuels for the entire energy system has disappeared, i.e., it is necessary to implement technologies that allow the generation of a gradual energy mix, which will make it possible to curb global warming and reduce dependence on fossil fuels in the world. Renewable energy technologies have positioned themselves in recent years as an alternative with projection and capacity, but with limitations due to their non-uniform use throughout the energy system, due to the variation in production. In addition, alternative technologies that can replace current industrial production techniques are in the process of centralized experimentation and have presented significant advances that provide an adequate future articulation to the world's energy supply [1].

On the other hand, carbon capture technology exists to contribute as much as renewable technologies in reducing the use of fossil fuels, however, carbon sequestration or capture is currently decentralized, with only a few applications at a centralized level, but it is expected that over the years, the technology will gain maturity and can contribute significantly to the current environmental problems.

It is important to mention that there are currently developing countries with unexplored fossil resources and a growing interest in using them to improve the quality of life of their inhabitants and to develop their economies. This scenario, in view of the current problems and the developed countries that have used their fossil resources for growth, generates an uncertain scenario with political tensions. However, a consensus must be generated so as not to slow down the growth of these developing countries and the appropriate use of renewable energies as an alternative for energy production in these countries.

Finally, the idea of developing smart energy grids that have common rules would provide a global opportunity for collaboration between technologies, generating a more efficient and sustainable consumption and generation system. Generating social and economic development worldwide. Integrating the rational use of fossil fuels and alternative technologies, which could gradually replace a large part of the consumption percentages of fossil fuels. It is important to link emerging economies and developing countries to change the dynamics of current policies and help form a strong global climate agreement.

3.2. Current Status of Projects in the World

Concentrating solar power is used worldwide in the industrial sector to supply heat or to produce electricity. Figure 3 presents the percentages of energy from solar collection in sectors such as: residential, industrial, transportation and others. The sector with the highest energy consumption is the industrial sector, followed by the transportation sector with 32% and 31%, respectively. The industrial sector of the 100% of energy consumed, only 26% is electrical energy, which shows a high demand for thermal consumption.

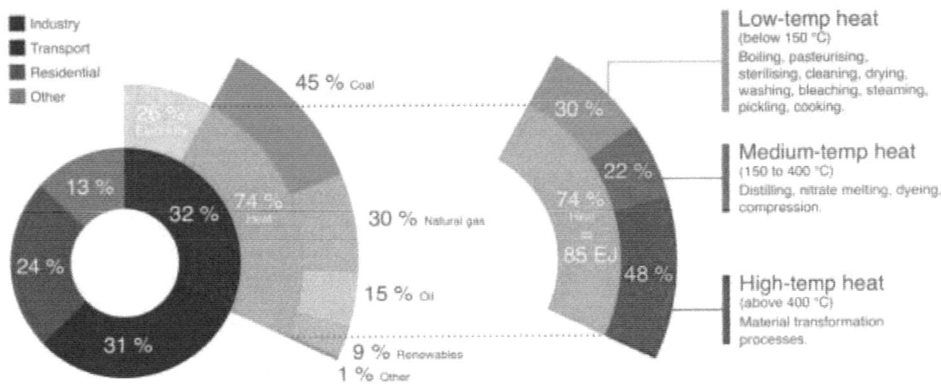

Figure 3. Use of solar thermal energy in consumer sectors [16]

The National Renewable Energy Laboratory (NREL) has a database of all existing centralized concentration projects, describing location, capacity, status (operational or not operational), among other relevant data. Additionally, it presents a mapping as shown in Figure 4. The 2D and 3D concentrating solar power systems are sometimes integrated with photovoltaic systems, however, in the compilation of information describing the existing concentrating solar power plants in the world, it was not possible to discriminate whether the parabolic trough systems were integrated with other systems (photovoltaic-PV).

Figure 4. Solar energy projects in the world [16]

3.3. Critical Literature Review

Concentrating solar power (CSP) has been in the research orbit recently, as it shows favorable environmental results compared to other technologies [17] [18] that use fossil fuels for their operation [19]. Concentration technologies are classified into 2D and 3D systems [20]. The 2D systems are technologically divided into Parabolic Trough Collectors (PTC) and Linear Fresnel Reflectors (LFR). 3D systems are technologically divided into parabolic dish collectors (PDC) and solar power towers (SPT) [21].

The two technologies that currently dominate the market are PTC and SPT, with parabolic trough collectors (PTC) being the most mature technology, followed by SPT, LFR and PDC respectively [22] [23]. The PTC and LFR systems are linear concentration collectors, while the SPT and PDC systems are point concentration. Linear concentration technologies have lower concentration ratios compared to point concentration technologies [24].

PTC, SPT and LFR technologies are recommended for power generation systems with sizes ranging from 10 to 200 MW, while PDC technology is recommended for low generation between 0.01 and 0.4 MW [25] [26]. Currently, there are about 135 CSP plants (see Table 6) operating in the world, 80% of which implement PTC technology, being the technology with the most operational and commercial experience at a global level. [27]. Consequently, PTC has the lowest technical and financial risks in the market, followed by SPT technology [28] [22].

On the other hand, the PDC technology, when implemented at a low scale, presents minimum soil occupation, while technologies such as PTC, LFR and SPT present high soil occupations, however, the PTC technology currently presents the best ratio of the three with respect to soil occupation and efficiency [29] [30]. This trend may tend to decrease due to the growth of SPT technology, however, it depends on market dynamics to gain more commercial experience in the coming years [31] [12].

Technologically, 2D or linear systems offer greater flexibility in construction, operation and maintenance [32] [33]. 3D technologies, on the other hand, are designed to generate more waste heat, implying a higher investment in the development of the systems [34] [35]. In addition, the life cycle of the 4 technologies is usually similar, depending on the size of the plant and the capacity, periodic intervention plans must be generated to guarantee the operability of the system [36] [37].

Finally, concentrating solar power technologies in general require identical weather conditions, with high normal direct solar irradiation (DNI) for long daily periods of the year, to ensure their operation and profitability [38] [39]. It is therefore recommended that for areas with extreme radiation levels such as coastal areas [40] [41].

Table 1 presents the matrix of the relationship between attributes and objectives; the attributes that meet the stated objectives are marked with an X.

Table 1. Technology Selection Matrix CPS

Objects / **Attributes**	**PTC**	**LFR**	**PDC**	**SPT**
A1	X		X	
A2	X	X		
A3	X			
A4	X	X	X	X
A5	X			X
A6	X	X	X	X
A7			X	X

4. Results

The current state of technology places Spain and the United States as world leaders in the process of generating electricity from concentrating solar power (CSP) plants, with the installed capacity in both countries exceeding 73% of the global CSP installation capacity in the world.

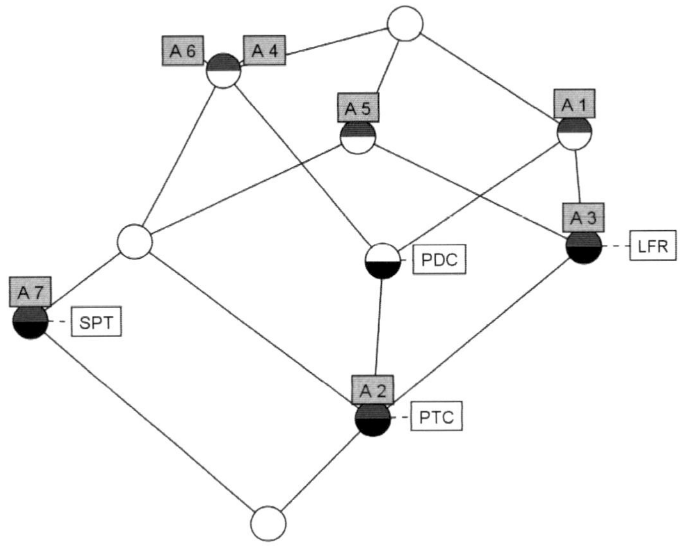

Figure 5. General FCA analysis diagram CSP Technologies

Considering the CSP projects under development worldwide, China is projected to be the country with the largest investment in concentrated solar power. China is currently building around 15 CSP plants for a total capacity of approximately 1064 MWe. The Chinese government has a vision towards renewable energies, ratified at COP21, which seeks to limit the accelerated process of global warming that the world is undergoing. Other developing countries such as Chile, India, the United Arab Emirates, Morocco and South Africa have developed CSP power generation projects, supported by various organizations such as the World Bank's Global Environment Facility, the German Federal Enterprise for International Cooperation and the Desertec Foundation. It is important to note that Chile is the first country in South America to have a centralized SPS system.

The results of the methodological application of the formal analysis of concepts through the Conception Explorer software are presented below. The different nodes shown in Figure 18 are identified in the following way:

- Blue-white: contains only attributes
- Blue-black: these are containers of solar tracking technologies related to attributes
- Black-White: Control Algorithm Only Containers
- White: Nodes acting as connectors between branches

Considering the above, the analysis of the diagram is simple. The four concentrating solar power (CSP) technologies are related to the attributes that, in turn, connect from the top down to the node of origin. At a glance, Figure 5 shows that attributes A1 "Cost", A2 "Technological complexity", A3 "Maturity", A4 "Life cycle", A5 "Implementation area", A6 "Weather conditions" and A7 "System efficiency" are not fully met by all objects. However, it is evident that attributes A4 and A6 are contained in all objects.

Figure 6 shows in blue the line that connects attributes with the specific objective of the PTC technology, highlighting:

- PTC technology meets attributes A1, A2, A3, A4, A5 and A6. PDC and TRF technologies are evident in the graph because they share attributes with PTC.
- The PTC technology within the present analysis is the technology with the greatest potential for implementation globally for the specific characteristics.

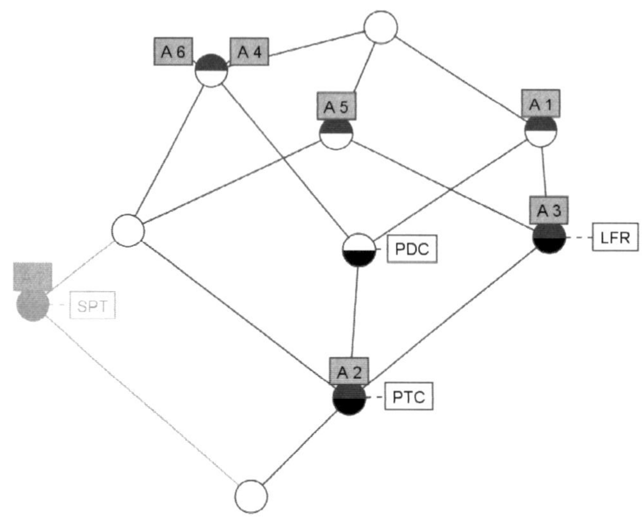

Figure 6. Specific FCA diagram for PTC technology

On the other hand, Figure 7 presents in blue color the line that connects attributes with the specific objective of the TRF technology, highlighting:

- The TRF technology complies with attributes A1, A2, A3 and A5. PTC technologies are evident in the graph because they share compliance with all the attributes mentioned above.
- The TRF technology within the present analysis is the technology with a medium global implementation potential for the specific characteristics evaluated.

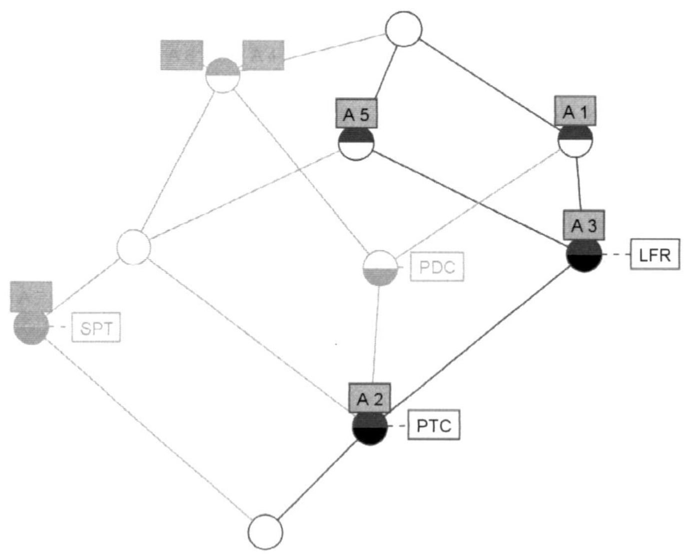

Figure 7. Specific FCA diagram for TRF technology

In turn, Figure 8 shows in blue the line that connects attributes with the specific objective of the SPT technology, highlighting:

- SPT technology complies with attributes A4, A5, A6 and A7. The graph does not show the other technologies; however, the white circles are nodes that link objects to each other, so it is evident that they share the fulfillment of all the attributes mentioned above.

IOP Conf. Series: Materials Science and Engineering 1299 (2024) 012009 doi:10.1088/1757-899X/1299/1/012009

- The SPT technology in the present analysis is the technology with a high global implementation potential for the specific characteristics evaluated.

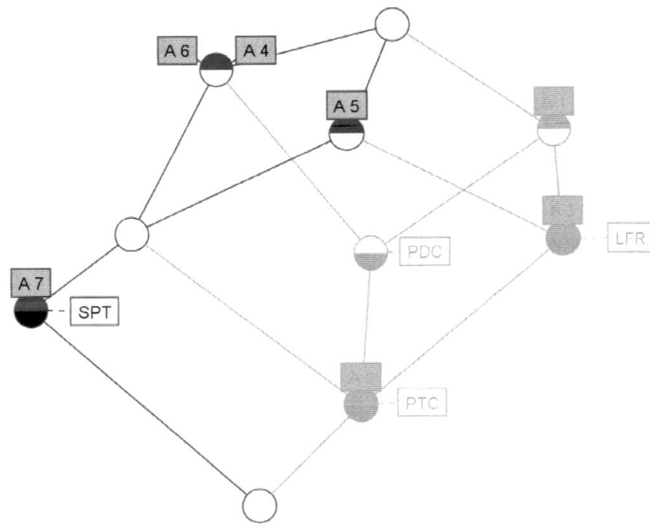

Figure 8. Diagrama especifico FCA para tecnología SPT

Additionally, Figure 9 presents in blue color the line that connects attributes with the specific objective of the PDC technology, highlighting:

- The PDC technology complies with attributes A1, A2, A4 and A6. PTC technologies are evident in the graph because they share compliance with all the attributes mentioned above.
- The PDC technology in the present analysis is the technology with a low global implementation potential for the specific characteristics evaluated.

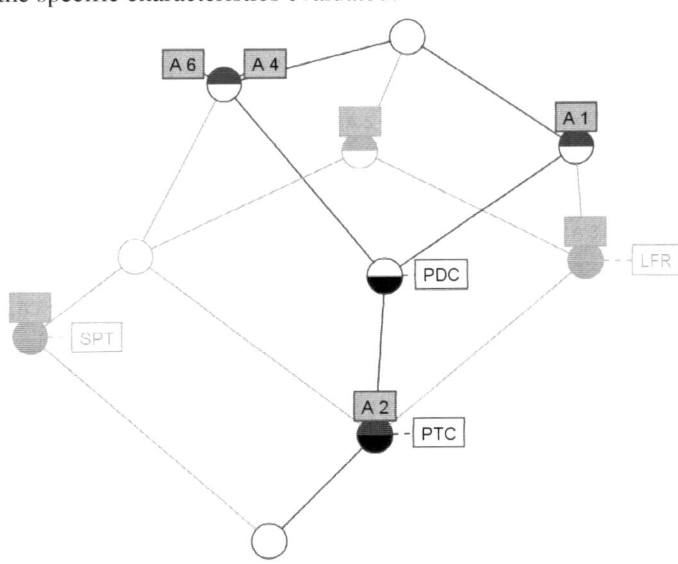

Figure 9. Specific FCA diagram for PDC technology

Finally, PTC technology is identified as the most suitable technology to be implemented in the future in Colombia based on the specific attributes proposed for the evaluation of the information.

5. Conclusions

The use of fossil fuels is one of the main causes of global warming, the excessive use of petroleum derivatives for the daily use of the human race has been excessive, which requires the use of alternative energy systems, being concentrating systems a promising technological line. However, it should be emphasized that the use of fossil fuels cannot be totally eliminated; rather, they should be gradually replaced over time, in order to generate a gradual decrease in global warming and a broader energy mix in the world.

Concentrating solar power (CSP) technologies are presented in the current world scenario as a key source of flexibility and application for electricity generation as an alternative to fossil fuels. To strengthen its implementation, a climate policy framework and a reduction in implementation costs are needed to generate greater impact at the centralized level.

CSP technologies face challenges for large-scale deployment that include lowering the cost compared to photovoltaic (PV) systems. At the same time, they require market mechanisms that guarantee the use of the energy generated by users, in order to maximize the economic contributions that would allow the growing development of more projects worldwide.

PTC technology is the technology with the greatest potential for implementation worldwide for centralized power generation systems. Based on the attributes defined to evaluate its potential application in future projects in Colombia, the PTC technology stood out for its cost-performance ratio, simple technology to implement, maintain and operate, as well as greater commercial maturity. These characteristics make it a promising technology in the process of applying new electric generation technologies in the country, generating an energy mix that feeds the national grid.

SPT technology is promising and it is expected that, over time, it could match or surpass PTC technology in terms of benefits. TRF technology is still in the process of commercial development, so the time to maturity may be longer than with SPT technology. Lastly, PDS systems are aimed at small generation plants, so their application in a centralized manner has been limited to date.

Finally, the application of the formal concept analysis (FCA) methodology allowed a critical evaluation of CSP technologies by directing the search for information to previously established parameters. Although the FCA methodology allows an interesting analysis of the information, it should not be replaced by numerical methods that allow information to be obtained with higher levels of precision; rather, it should be part of a first stage of literature review.

References

[1] S. Foster y D. Elzinga, «El papel de los combustibles fósiles en un sistema energético sostenible | Naciones Unidas», *United Nations*, United Nations, 2021. Accedido: 21 de enero de 2022. [En línea]. Disponible en: https://www.un.org/es/chronicle/article/el-papel-de-los-combustibles-fosiles-en-un-sistema-energetico-sostenible

[2] C. Garrett, «El calentamiento global en 2022: causas y consecuencias», *Selectra*, 11 de febrero de 2021. https://climate.selectra.com/es/que-es/calentamiento-global (accedido 21 de enero de 2022).

[3] ONU, «Renewable 2021 Global Status Report». 2021. Accedido: 7 de enero de 2022. [En línea]. Disponible en: https://www.ren21.net/wp-content/uploads/2019/05/GSR2021_Full_Report.pdf

[4] B. E. Tarazona-Romero, Y. A. M. Maldonado, A. C. Celador, y O. L. Pérez, «Optical Performance Assessment of a Handmade Prototype of Linear Fresnel Concentrator», *Periodicals of Engineering and Natural Sciences*, vol. 9, n.º 4, Art. n.º 4, oct. 2021, doi: 10.21533/pen.v9i4.1987.

[5] J. J. Delgado Torres, «Energía renovables en el Latinoamérica, aplicabilidad y proyección en Colombia: Revisión», oct. 2021, Accedido: 21 de enero de 2022. [En línea]. Disponible en: http://repositorio.uts.edu.co:8080/xmlui/handle/123456789/7581

[6] J. A. Roca, «Las 10 plantas termosolares más grandes del mundo: EEUU y España se reparten el 90% del pastel», 2020. https://elperiodicodelaenergia.com/las-10-plantas-termosolares-mas-grandes-del-mundo-eeuu-y-espana-se-reparten-el-90-del-pastel/ (accedido 21 de enero de 2022).

[7] B. Ganter y R. Wille, *Análisis de concepto formal, Texto original: Fundamentos matemáticos*. Berlín Heidelberg: Springer-Verlag, 1999. Accedido: 9 de julio de 2021. [En línea]. Disponible en: https://www.springer.com/gp/book/9783540627715
[8] F. Sancho Caparrini, «Análisis Formal de Conceptos», *Análisis Formal de Conceptos*, 2017. http://www.cs.us.es/~fsancho/?e=78 (accedido 8 de julio de 2021).
[9] S. Yevtushenko, J. Tane, T. B. Kaiser, S. Objedkov, J. Hereth Correia, y H. Reppe, *Concept Explorer*. 2021. [En línea]. Disponible en: http://conexp.sourceforge.net/download.html
[10] M. Thompson, «Wicked environmental problems», en *Companion to Environmental Studies*, Routledge, 2018.
[11] M. B. Fayzullayevich, «The Problem of Environmental Problems in the Regions of the World», *International Journal of Innovative Analyses and Emerging Technology*, vol. 1, n.º 6, Art. n.º 6, nov. 2021.
[12] S. Arsovski, M. Kwiatkowski, A. Lewandowska, D. J. Peshevska, E. Sofeska, y M. Dymitrow, *Can urban environmental problems be overcome? The case of Skopje—world's most polluted city*. Wydawnictwo Naukowe Uniwersytetu Mikołaja Kopernika, 2018. Accedido: 8 de enero de 2022. [En línea]. Disponible en: https://www.ceeol.com/search/article-detail?id=738098
[13] ONU, «Cambio climático | Naciones Unidas», *United Nations*, 2022. https://www.un.org/es/global-issues/climate-change (accedido 22 de enero de 2022).
[14] V. Masson-Delmotte *et al.*, Eds., *Climate Change 2021: The Physical Science Basis. Contribution of Working Group I to the Sixth Assessment Report of the Intergovernmental Panel on Climate Change*. Cambridge University Press, 2021.
[15] D. Flórez-Orrego, R. A. Freire, J. A. M. da Silva, C. Albuquerque Neto, y S. de Oliveira Junior, «Centralized power generation with carbon capture on decommissioned offshore petroleum platforms», *Energy Conversion and Management*, vol. 252, p. 115110, ene. 2022, doi: 10.1016/j.enconman.2021.115110.
[16] NREL, «Concentrating Solar Power Projects by Country», *Solar Space NREL*, 2021. https://solarpaces.nrel.gov/concentrating-solar-power-projects-country (accedido 27 de enero de 2022).
[17] U. Desideri, F. Zepparelli, V. Morettini, y E. Garroni, «Comparative analysis of concentrating solar power and photovoltaic technologies: Technical and environmental evaluations», *Applied Energy*, vol. 102, pp. 765-784, feb. 2013, doi: 10.1016/j.apenergy.2012.08.033.
[18] L. Xu *et al.*, «Analysis of optical and thermal factors' effects on the transient performance of parabolic trough solar collectors», *Solar Energy*, vol. 179, pp. 195-209, feb. 2019, doi: 10.1016/j.solener.2018.12.070.
[19] R. Wang, J. Sun, y H. Hong, «Proposal of solar-aided coal-fired power generation system with direct steam generation and active composite sun-tracking», *Renewable Energy*, vol. 141, pp. 596-612, oct. 2019, doi: 10.1016/j.renene.2019.04.013.
[20] G. Morin, «Chapter 15 - Optimization of concentrating solar power plant designs through integrated techno-economic modelling», en *Concentrating Solar Power Technology (Second Edition)*, K. Lovegrove y W. Stein, Eds. Woodhead Publishing, 2021, pp. 545-583. doi: 10.1016/B978-0-12-819970-1.00010-4.
[21] K. S. Reddy y C. Ananthsornaraj, «Design, development and performance investigation of solar Parabolic Trough Collector for large-scale solar power plants», *Renewable Energy*, vol. 146, pp. 1943-1957, feb. 2020, doi: 10.1016/j.renene.2019.07.158.
[22] O. Achkari y A. El Fadar, «Latest developments on TES and CSP technologies – Energy and environmental issues, applications and research trends», *Applied Thermal Engineering*, vol. 167, p. 114806, feb. 2020, doi: 10.1016/j.applthermaleng.2019.114806.
[23] B. E. Tarazona-Romero, A. C. Celador, C. L. S. Rodriguez, J. G. A. Villabona, y A. D. R. Quintero, «Design and construction of a solar tracking system for Linear Fresnel Concentrator», *Periodicals of Engineering and Natural Sciences*, vol. 9, n.º 4, Art. n.º 4, oct. 2021, doi: 10.21533/pen.v9i4.1988.

[24] W.-D. Steinmann, «Chapter 11 - Thermal energy storage systems for concentrating solar power plants», en *Concentrating Solar Power Technology (Second Edition)*, K. Lovegrove y W. Stein, Eds. Woodhead Publishing, 2021, pp. 399-440. doi: 10.1016/B978-0-12-819970-1.00008-6.
[25] B. Belgasim, Y. Aldali, M. J. R. Abdunnabi, G. Hashem, y K. Hossin, «The potential of concentrating solar power (CSP) for electricity generation in Libya», *Renewable and Sustainable Energy Reviews*, vol. 90, pp. 1-15, jul. 2018, doi: 10.1016/j.rser.2018.03.045.
[26] R. Meyer, M. Schlecht, K. Chhatbar, y S. Weber, «Chapter 3 - Solar resources for concentrating solar power systems», en *Concentrating Solar Power Technology (Second Edition)*, K. Lovegrove y W. Stein, Eds. Woodhead Publishing, 2021, pp. 73-98. doi: 10.1016/B978-0-12-819970-1.00014-1.
[27] E. Z. Moya, «Chapter 7 - Parabolic-trough concentrating solar power systems», en *Concentrating Solar Power Technology (Second Edition)*, K. Lovegrove y W. Stein, Eds. Woodhead Publishing, 2021, pp. 219-266. doi: 10.1016/B978-0-12-819970-1.00009-8.
[28] C. Turchi, M. Mehos, C. K. Ho, y G. J. Kolb, «Current and Future Costs for Parabolic Trough and Power Tower Systems in the US Market: Preprint», National Renewable Energy Lab. (NREL), Golden, CO (United States), NREL/CP-5500-49303, oct. 2010. Accedido: 27 de enero de 2022. [En línea]. Disponible en: https://www.osti.gov/biblio/991556
[29] M. T. Islam, N. Huda, A. B. Abdullah, y R. Saidur, «A comprehensive review of state-of-the-art concentrating solar power (CSP) technologies: Current status and research trends», *Renewable and Sustainable Energy Reviews*, vol. 91, pp. 987-1018, ago. 2018, doi: 10.1016/j.rser.2018.04.097.
[30] B. E. Tarazona-Romero, A. Campos-Celador, Y. A. Muñoz-Maldonado, J. G. Ascanio-Villabona, M. A. Duran-Sarmiento, y A. D. Rincón-Quintero, «Development of a Fresnel Artisanal System for the Production of Hot Water or Steam», en *Recent Advances in Electrical Engineering, Electronics and Energy*, Cham, 2021, pp. 196-209. doi: 10.1007/978-3-030-72212-8_15.
[31] S. Izquierdo, C. Montañés, C. Dopazo, y N. Fueyo, «Analysis of CSP plants for the definition of energy policies: The influence on electricity cost of solar multiples, capacity factors and energy storage», *Energy Policy*, vol. 38, n.º 10, pp. 6215-6221, oct. 2010, doi: 10.1016/j.enpol.2010.06.009.
[32] H. Price *et al.*, «Chapter 20 - Concentrating solar power best practices», en *Concentrating Solar Power Technology (Second Edition)*, K. Lovegrove y W. Stein, Eds. Woodhead Publishing, 2021, pp. 725-757. doi: 10.1016/B978-0-12-819970-1.00020-7.
[33] L. Crespo, «Chapter 13 - The long-term market potential of concentrating solar power systems», en *Concentrating Solar Power Technology (Second Edition)*, K. Lovegrove y W. Stein, Eds. Woodhead Publishing, 2021, pp. 477-509. doi: 10.1016/B978-0-12-819970-1.00004-9.
[34] M. McPherson, M. Mehos, y P. Denholm, «Leveraging concentrating solar power plant dispatchability: A review of the impacts of global market structures and policy», *Energy Policy*, vol. 139, p. 111335, abr. 2020, doi: 10.1016/j.enpol.2020.111335.
[35] B. E. Tarazona-Romero, A. Campos-Celador, y Y. A. Maldonado-Muñoz, «Can solar desalination be small and beautiful? A critical review of existing technology under the appropriate technology paradigm», *Energy Research & Social Science*, vol. 88, p. 102510, jun. 2022, doi: 10.1016/j.erss.2022.102510.
[36] G. Brumana, G. Franchini, y E. Ghirardi, «Investigation of the Load-Following Capability of CSP Plants», *Energy Procedia*, vol. 148, pp. 615-622, ago. 2018, doi: 10.1016/j.egypro.2018.08.149.
[37] K. Lovegrove y J. Pye, «Chapter 2 - Fundamental principles of concentrating solar power systems», en *Concentrating Solar Power Technology (Second Edition)*, K. Lovegrove y W. Stein, Eds. Woodhead Publishing, 2021, pp. 19-71. doi: 10.1016/B978-0-12-819970-1.00013-X.
[38] O. Ogunmodimu y E. C. Okoroigwe, «Concentrating solar power technologies for solar thermal grid electricity in Nigeria: A review», *Renewable and Sustainable Energy Reviews*, vol. 90, pp. 104-119, jul. 2018, doi: 10.1016/j.rser.2018.03.029.

[39] W. J. Platzer y C. Hildebrandt, «Chapter 14 - Absorber materials for solar thermal receivers in concentrating solar power systems», en *Concentrating Solar Power Technology (Second Edition)*, K. Lovegrove y W. Stein, Eds. Woodhead Publishing, 2021, pp. 511-544. doi: 10.1016/B978-0-12-819970-1.00001-3.
[40] S. Ravelli, G. Franchini, y A. Perdichizzi, «Comparison of different CSP technologies for combined power and cooling production», *Renewable Energy*, vol. 121, pp. 712-721, jun. 2018, doi: 10.1016/j.renene.2018.01.074.
[41] W.-D. Steinmann y C. Prieto, «24 - Thermal storage for concentrating solar power plants», en *Advances in Thermal Energy Storage Systems (Second Edition)*, L. F. Cabeza, Ed. Woodhead Publishing, 2021, pp. 673-697. doi: 10.1016/B978-0-12-819885-8.00024-3.

IOP Conf. Series: Materials Science and Engineering 1299 (2024) 012010 doi:10.1088/1757-899X/1299/1/012010

Comparative analysis of the current energy of Colombia and the countries of its meridian.

J G Ascanio-Villabona[1], O Lengerke-Pérez[1], B E Tarazona-Romero[1], K T Jaimes-Quintero[1] , C L Sandoval-Rodríguez and M A Duran-Sarmiento[1].
[1] Unidades Tecnológicas de Santander UTS, Bucaramanga Santander 680005, Colombia.
Jgascanio@correo.uts.edu.co, Olenguerke@correo.uts.edu.co, Btarazona@correo.uts.edu.co, Ktjaimes@uts.edu.co, Csandoval@correo.uts.edu.co, and Aduran@correo.uts.edu.co.

Abstract. The research focuses on the evaluation of the progress of the renewable installed capacity of the meridian countries with Colombia. Junto a Colombia, there are 12 other countries that are crossed by an imaginary line that surrounds the planet and because of this, each of these countries that it crosses, have a similar climatology, which, compared to Colombia, is assimilated to have the same conditions for the implementation of renewable energies. A bibliographic compilation was carried out where the amount of published information that can show the moment that the energy sector of each country that belongs to the Equatorial Meridian is going through was taken into account, with the aim of analyzing and comparing the energy news of Colombia with respect to these. This study shows the current energy state of each country, where different points are observed in relation to renewable and conventional energies. This comparative analysis was carried out based on the information found and gives a good impression of Colombia in the current and future energy field, but with difficulties due to factors outside its projection.

1. Introduction

Among the challenges for the XXI century is energy consumption and the impact for climate change [1]. Colombia has focused on improvement systems for the energy sector with sustainable energy planning strategies implementing sustainable energy sources and their challenges associated with obtaining and distributing them. [2]. Through studies and research, it has been possible to carry out some of these projects based on energy systems similar to those proposed and with this, create a more environmentally friendly development, trying to improve the demand for energy [2]–[4].

The transition to this type of unconventional energies are not only on the part of Colombia, but are also the present of the vast majority of countries in the world [5]–[7], since, thanks to the use of these energy sources, costs can be reduced and economic benefits obtained [8]. In order to perceive the behavior in the energy sector in Colombia, a comparative analysis is carried out based on simular climatological zones [9], so for this study, it takes into account the countries that cross the line of the Equatorial meridian as that Colombia, which allows, to observe their development with respect to these countries that have an equal or similar climate, so the methods to take advantage of these unconventional energies have a factor in common and with this a real improvement can be shown or on the contrary a lack of energy development[10].

2. Literature review

For the bibliometric review, the analysis of the current state and conventional and unconventional energy trends in Colombia is carried out. When searching with keywords such as "conventional" and "unconventional" in Colombia, 7901 document results are found. For this first search, the , is obtained, where the visualization of the network of the keywords related to the theme is presented **Figure *1*** network display of related keywords

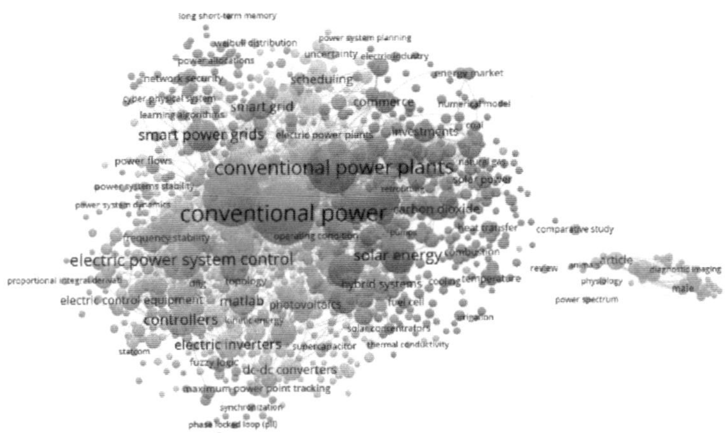

Figure 1 network display of related keywords

Aitself the discrimination of the words that are indexed in scopus is carried out by performing an analysis of co-occurrence of indexed words as presented in Fig. 2.

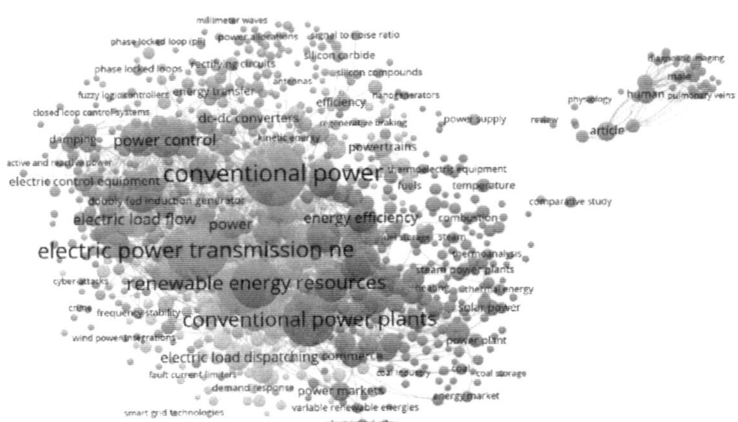

Figure 2 co-occurrence overlap of indexed words

In the Table 1 the 15 words with total force mass of link between the different indexed words and the occurrence of the same are presented.

IOP Conf. Series: Materials Science and Engineering 1299 (2024) 012010 doi:10.1088/1757-899X/1299/1/012010

Table 1 Co-occurrence of indexed words

N°	keyword	occurrences	total link strength
1	conventional power	424	2918
2	electric power transmission networks	375	3778
3	conventional power plants	277	2611
4	renewable energy resources	241	2531
5	wind power	236	2294
6	electric power system control	171	1893
7	renewable energy source	171	1849
8	solar energy	123	1228
9	electric load flow	120	1140
10	smart power grids	115	1131
11	solar power generation	114	1207
12	renewable energies	111	1099
13	power control	106	1042
14	controllers	104	1121
15	energy storage	94	989

Continuing with the documentary purification, 300 articles are discriminated against, in this search those articles that are related to the topic of the current energy of Colombia and the countries of the Equator line were taken into account, so they select 28 articles with which it was developed in the comparison based on 26 articles that provide information about the energies (whether conventional or unconventional) that are used or that are take into account to implement over the next few years in order to obtain an improvement in the energy system of each country. In the Table 2 you can see the 24 journals in which the consulted articles that were chosen and those collected related to the energy present of Colombia and the countries that share its meridian were published..

Table 2 Selected journals and bibliographic material.

Selected journals	Bibliographic material consulted	Bibliographic material collected
AIMS Energy	1	1
Fuel Communications	1	1
FISE-IEEE/CIGRE Conference 2019 - Living the Energy Transition (FISE/CIGRE)	1	1
Ocean Thermal Energy Conversion (OTEC): Past, Present, and Progress	1	1
The Asia Journal Finance, Economics and Business	1	1
Energy Efficiency	1	1
Energy	1	1
Energies	2	2
Applied Energy	1	1
Renewable energy	1	1
Energy for sustainable development	1	1
Focus on renewable energy	1	1
Environmental Innovation and Social Transitions	1	0
Energy policy	3	2
Biomass burning in Africa	1	1
Renewable & Sustainable Energy Reviews	1	1
Energy Strategy Reviews	1	1
Journal of Research & Energy Reviews	1	1
Cleaner Production Magazine	2	2
International Journal of Energy Economics and Policy	1	1
Sustainability	1	1
University Science and Technology	1	1
2019 8th International Conference on Systems and Control (ICSC)	1	1
2020 Tenth Seminar on Electrical Energy, Electronics, Communications, Controls and Informatics (EECCIS)	1	1
TOTAL	28	26

2.1. Comparative studies

The material was collected with the criterion of high degree of topicality, so relevant information is determined with a maximum of 4 years old, so only the information published in a period of time between 2019 and 2021 is taken into account. Due to the little information found about the current energy situation of each country involved in the preparation of this document, there are projects, research, reports and studies that address the current energy issue and future projections, with which an idea of the progress of each of these countries can be evidenced and thus be able to analyze the present of Colombia and the other countries belonging to the Equator line. (See Table 3).

IOP Conf. Series: Materials Science and Engineering 1299 (2024) 012010 doi:10.1088/1757-899X/1299/1/012010

Table 3 *List of Selected Articles*

N°	Año	Titulo
1	2019	Distributed generation: an inventory of installed capacity in Colombia until 2019[11]
2	2019	Renewable energy in Colombia and the opportunity for offshore wind technology [12]
3	2019	Large-scale integration of renewable energy sources (RES) into the future Colombian energy system[3]
4	2019	A look at electricity generation from non-conventional renewables in Colombia [13]
5	2020	The energy potential of agricultural, agro-industrial, livestock and slaughterhouse biomass waste through direct combustion and anaerobic digestion. The case of Colombia [14]
6	2020	The role of renewable energy for sustainable energy governance and environmental policies for climate change mitigation in Ecuador [15]
7	2019	Time series analysis to predict hydroelectric power production: the case of Ecuador [16]
8	2020	The use of hydropower for electric power generation as a strategy for industrial development in Ecuador [17]
9	2019	Big Oil's Renewable Energy Strategies: From Oil to Energy? [18]
10	2019	Ocean Renewable Energy Potential, Technology and Deployments: A Case Study from Brazil[19]
11	2020	Towards a smart grid electricity system in Brazil: challenges and opportunities [20]
12	2020	Biomass burning in the Democratic Republic of the Congo [21]
13	**2019**	A comprehensive review on the state of solar PV growth in Uganda [22]
14	2019	A long-term, multi-criteria energy planning optimization model with on- and off-grid electrification: the case of Uganda [23]
15	**2019**	A comprehensive review of the energy scenario and sustainable energy in Kenya[24]
16	**2019**	Decentralised rural electrification in Kenya: accelerating universal access to energy [25]
17	**2019**	Review of Solar Energy Development in Kenya: Opportunities and Challenges [26]]
18	**2020**	Electrificación rural en Kenia: un caso útil para áreas remotas en África subsahariana [27]
19	**2019**	Powering an island complex with renewable energy: a feasibility analysis in the Maldives [28]
20	**2020**	Resource efficiency in the electricity sector in the Maldives [29]
21	**2019**	Waste-to-energy technology: the potential of sustainable biogas production from animal waste in Indonesia [30]
22	**2019**	The state of energy poverty in Indonesia and its impact on well-being [31]
23	**2020**	Policy research to improve the adoption of rooftop home PV systems in Indonesia [32]
24	**2020**	Biodiesel produced from palm oil in Indonesia: current status and opportunities [33]
25	**2020**	Optimized renewable energy design, firm capacity and stable operating power plant to serve offgrid commercial loads in Indonesia [34]
26	**2020**	Can ocean thermal energy conversion and seawater use help small island developing States? A case study from Kiribati, Pacific Islands region[35

Based on the total bibliographic material consulted and collected, there were countries such as Sao Tome and Principe, Gabon, Republic of the Congo and Somalia, in which relevant information related to the current energy situation of these countries was not obtained and therefore were not taken into account when developing the document. Based on the above, it can be evidenced the amount of renewable energy that has been installed until 2020 in MegaWatts (MW). In the Table 4 se la capacidad instalada por país y por sistema de generación implementada.

Table 4 Sustainable energy capacity 2020 (MW)

Democratic Republic of the Congo	19	0	2750	3	0	
Uganda	82	0	907	88	0	1077
Kenya	95	336	837	88	823	2179
Maldives	14	1	0	0	0	15
		Wind	**Hydroelectric**	**Bioenergy**	**Geotermal**	**Total**
Colombia	90	**18**	**12258**	**340**	**0**	**12706**
Ecuador	28	**21**	**5079**	**152**	**0**	**5280**
Brazil	2485	**15364**	**109092**	**14992**	**0**	**141933**
						2772

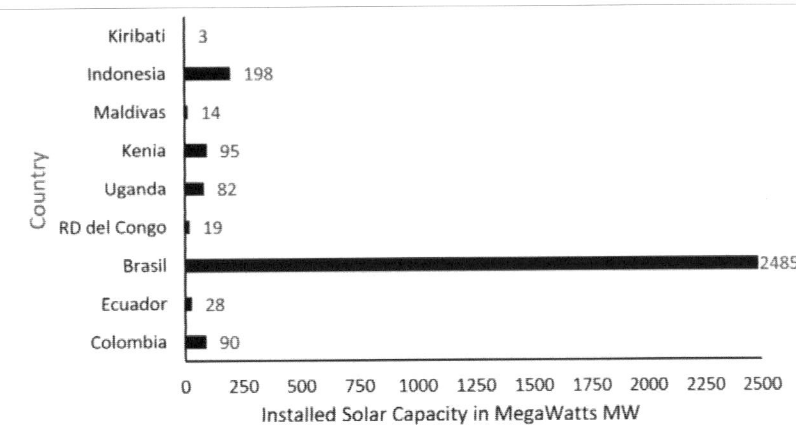

Figure 3 Solar Capacity Installed in MegaWatts (MW)

In relation to renewable sources, Colombia is very weak since as illustrated inFigure 3, the installed solar energy capacity is 90MW which is equivalent to 0.7% of the total installed capacity in Colombia and is below countries such as Brazil, Indonesia and Kenya who have so far had greater projection when it comes to obtaining energy this type of sustainable source. With respect to wind energy, like solar energy, the development and projection of this type of renewable energy source is very low, since its installed capacity of 18MW, equivalent to only 0.14% of the total installed capacity and, in addition, it is below the vast majority of countries that share its meridian. On the other hand, inFigure 4, it can be seen that the use of hydroelectric energy sources is greater than the vast majority of countries with the exception of Brazil, where a positive advance can be idealized. This type of renewable energy is the most used in Colombia, with a capacity of 12258MW, which is equivalent to 96.47% of all installed renewable energies in the country.

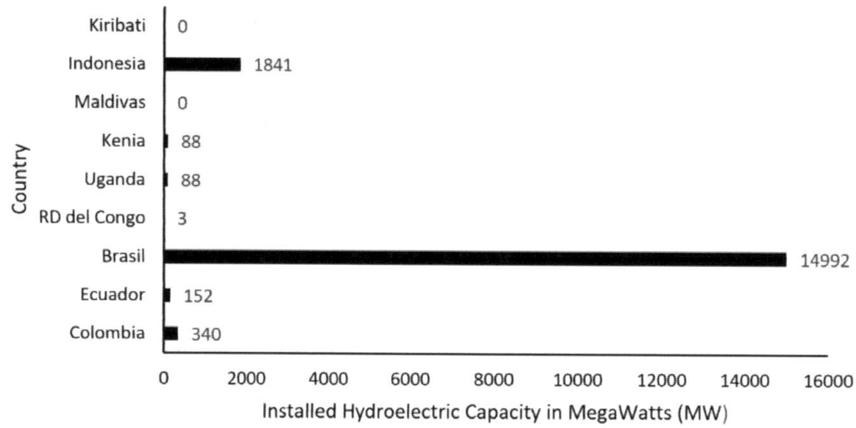

Figure 4 Installed Hydroelectric Capacity in MegaWatts (MW)

2.2. Characterization of Colombia's energy news

Colombia is in the process of development and improvement, according to the bibliographic material collected rebuked in the , the exploitation of conventional energies that are used until now, so like the other countries that crosses the line of the Equator maintain an idea of making an energy transition towards unconventional sources that allow benefits in the energy sector, that goes hand in hand with environmental, economic, social, climatological sectors, among others. In addition to getting energy to as many sectors as possible for better access to this resource.

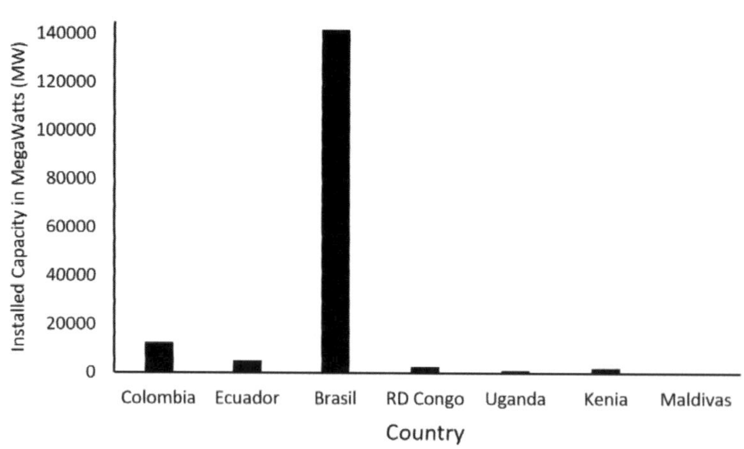

Figure 5 Total installed renewable energy capacity

The totality of Mega-Watts that are installed in each of the countries that are crossed by the Equator, can be reflected in Figure 5 Total installed renewable energy capacity. Colombia's energy transition to a more sustainable and environmentally friendly country is in process. Currently Colombia is projecting positively towards an energy system that benefits the inhabitants, commercial areas, industries and all other sectors that require the use of energy for its operation, which benefits the largest population and sectors such as environmental, political, social, geographical, economic, among others that are directly and indirectly involved with the development of energy. Clean energy in Colombia has been implemented for many years to make a transition where there is less and less need to use conventional energies that are exhausted over time and damage the ecosystem. The amount of renewable resources that the country has can be used to the maximum so that this transition is as soon as possible, where most of the renewable energy that is being used is hydroelectric energy (See Figure 6), which benefits from the abundance of hydrographic sources found throughout the Colombian country.Figure 6 Energy mix of different energy sources5

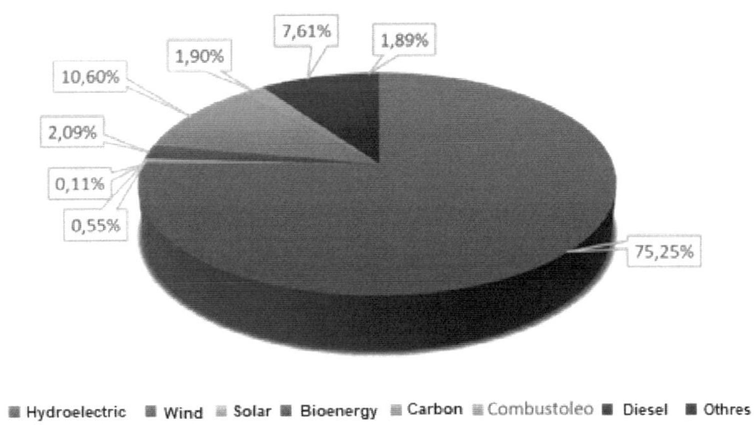

Figure 6 Energy mix of different energy sources5

3. Energy potential

3.1. wind energy

The energy potential of wind sources that Colombia has is mostly concentrated in the north of the country with approximately 27,000MW distributed in different regions as can be seen in Table 5, where a decrease between 15% and 35% can be seen as this potential is further investigated[13].

IOP Conf. Series: Materials Science and Engineering 1299 (2024) 012010 doi:10.1088/1757-899X/1299/1/012010

Table 5 Wind energy potential by region

Region	Potential energy capacity
Guajira Peninsula	18.000 MW
Santander	5.000 MW
North coast	1.000 MW
Other regions	3.000 MW
TOTAL	27.000MW

Some authors such as the case of Pupo Roncallo and his team cite that the wind potential could be approximately 81.2 TWh that can represent the installation of a capacity up to 25 Giga-Watts (GW). Likewise, wind energy can experience high growth if the projects that have been approved for a next execution are taken into account and thus provide a greater capacity than is installed, since Colombia is supplied with a single installed wind farm (Jeripachi project) for the generation of this type of energy [2], [3].

3.2. Potential in solar energy

On the other hand, thanks to the average solar radiation in Colombia dating between 4.5 to 6.0 kWh /^{m2}, there is a high potential for the use of photovoltaic solar technologies [3]. This radiation is above the world radiation average of 3.9 kWh / m^2, mainly due to the tropical climate that Colombia has when it is crossed by the Equator line and not having the 4 stations like the other countries far from the Equator. This potential is foreseen by the UPME for 2025 an increase between 148 and 168 MW and for 2030 between 163 and 216 MW, which can be considered low for a potential in solar radiation as high as that of this type of renewable source in Colombia [12], [13].

3.3. Potential in hydropower

Considering that the hydrographic sources available to the country are the most used to obtain energy, the development of this type of energy and the use of its potential has been of great importance in the country. The potential hydroelectric capacity that Colombia can count on is approximately 93 GigaWatts (GW), however, it is not a full and defined potential since there are many environmental factors that can limit it. To date, the installed capacity of 12,258 MW that is projected to continue to increase over the years [2], [3].

3.4. Bioenergy potential

The bioenergy available mainly from biomass between agricultural waste, forest residues, energy crops, residual biomass from livestock farms and biomass from municipal solid waste, can represent a high potential in this type of renewable energy with around 110,000 MW and 200,000 MW where 16,000MW could be used. So more research is needed to adapt this type of renewable sources and make the most of the energy that can be obtained from them[12], [14], [36]. There is also a vast untapped bioenergy in Colombia of approximately 116 TWh per year. Where with its use it could also generate and promote a development in the rural sector that diversifies the country's energy mix and there is greater sustainability leaving behind conventional energy sources. In general, the potential of the Colombian territory to obtain energy from renewable sources is high and with a great projection, so that, if each of these sources is fully used, it will be possible to count on large amounts of energy that benefit both the energy sector and the economic sector, social and environmental, with a strong and positive development that makes Colombia a sustainable country with better conditions in all possible aspects.

3.5. Factors that hinder good progress

Colombia is a country that has great energy potential, due to its diversity in resources to obtain energy. Unfortunately, there are cases in energy planning that miss all this potential, so Martínez and Castillo express that:

> *"The analysis of the Colombian case shows that the central objective of energy planning and its institutionality has been the advancement and maximization of the exploitation of energy resources for export purposes, instead of providing a guide to reorganize the system towards the reduction of energy consumption, non-renewable resources in favor of renewable sources, or energy justice"* [37].

According to the UPME energy exports and imports indicator [38], in 2020, 250,795 GWh were exported and in addition to this, the balance with respect to the energy imported by the country is negative, since in 2020 1,301,589 GWh were imported, which on the one hand, increases the internal energy cost of the country and on the other, the exported energy is wasted that would reduce the amount of energy imported. Another problem that the Colombian country presents is the little use of its wealth in renewable resources that are available due to the lack of research, investment and use of other sources other than water. For this reason, renewable energies are currently at a low level of development, which does not allow a wide repertoire of alternatives that are more geographically accessible as in the case of the department of La Guajira, where the difficulty in obtaining energy from water sources is great, and photovoltaic and wind energies that have a potential that can supply a large amount of energy demand in the country are more viable [39]. On the other hand, Colombia has also been affected by factors that, on the one hand, are difficult to control such as mega-droughts, which occur at any time and, on the other hand, demonstrate the absence of control in forested areas, since, due to the deforestation of forests by unknown external agents, increases the rate of possible droughts that affect the country's energy sector [40].

4. Conclusions

The energy sector does not depend on itself, because it goes hand in hand with other sectors such as economic, environmental, social, climatological, among others. Because of this dependence, the availability of bibliographic material at the time of research becomes scarce, since by sectors which require an approach according to the problems presented by each country, energy goes to the background becoming a topic of minor importance, which makes it difficult to obtain updated bibliographic material that is useful for the development of documents whose theme focuses on energy. For this reason, for the comparison between Colombia and the other countries that share its meridian, it was not possible to take into account all of the 13 countries involved in this issue, therefore, the comparison occurred only with those countries where they had available the most up-to-date material possible. With this bibliographic material, it was possible to observe each of the types of renewable energy sources that are used and the capacity of Mega-Watts (MW) that has been installed in countries whose climate is similar to that of Colombia. In the comparison made with the data obtained, there is a lack of renewable energy development in Colombia, since it is below several countries when it comes to taking advantage of renewable energy sources other than hydropower. There is also a trend of superiority on the part of Brazil, where its development has been very broad compared to Colombia and the rest of the countries, so it can be a model towards a future projection, to be able to exploit the energy potential that Colombia has, and to be able to compete in this sector in a more optimal way and with a more significant advance.

The transition to a sustainable energy country as it has been planning for the last few years, so far is slow and lacking investment in the various sustainable energy opportunities offered by the Colombian territory, therefore, the projects that are in the process of research and implementation to achieve this objective that the country has set are good. The progress that the energy sector has had in Colombia may become backward and insignificant, but taking full advantage of the renewable resources available to the country, the research that has been done over the years and the projects that are planned and are being implemented, this transition can be made in a faster way and with greater projection, for a Colombia that is more friendly to the environment and with greater accessibility in the regions that can currently benefit from this resource as important as energy..

5. References

[1] T. Hwang, S. Kang, and J. T. Kim, "Optimization of the building integrated photovoltaic system in office buildings—Focus on the orientation, inclined angle and installed area," *Energy Build.*, vol. 46, pp. 92–104, Mar. 2012, doi: 10.1016/J.ENBUILD.2011.10.041.

[2] O. Pupo-Roncallo, J. Campillo, D. Ingham, K. Hughes, and M. Pourkashanian, "Renewable energy production and demand dataset for the energy system of Colombia," *Data Br.*, vol. 28, p. 105084, Feb. 2020, doi: 10.1016/J.DIB.2019.105084.

[3] O. Pupo-Roncallo, J. Campillo, D. Ingham, K. Hughes, and M. Pourkashanian, "Large scale integration of renewable energy sources (RES) in the future Colombian energy system," *Energy*, vol. 186, p. 115805, Nov. 2019, doi: 10.1016/J.ENERGY.2019.07.135.

[4] A. R. López *et al.*, "Solar PV generation in Colombia - A qualitative and quantitative approach to analyze the potential of solar energy market," *Renew. Energy*, vol. 148, pp. 1266–1279, Apr. 2020, doi: 10.1016/J.RENENE.2019.10.066.

[5] R. Weijermars, G. Drijkoningen, T. J. Heimovaara, E. S. J. Rudolph, G. J. Weltje, and K. H. A. A. Wolf, "Unconventional gas research initiative for clean energy transition in Europe," *J. Nat. Gas Sci. Eng.*, vol. 3, no. 2, pp. 402–412, May 2011, doi: 10.1016/J.JNGSE.2011.04.002.

[6] K. Francis, S. Dongying, A. Dennis, N. N. K. Edmund, and N. Y. G. Kumah, "Network governance and renewable energy transition in sub-Saharan Africa: Contextual evidence from Ghana," *Energy Sustain. Dev.*, vol. 69, pp. 202–210, Aug. 2022, doi: 10.1016/J.ESD.2022.06.009.

[7] M. Tutak and J. Brodny, "Renewable energy consumption in economic sectors in the EU-27. The impact on economics, environment and conventional energy sources. A 20-year perspective," *J. Clean. Prod.*, vol. 345, p. 131076, Apr. 2022, doi: 10.1016/J.JCLEPRO.2022.131076.

[8] D. Cielo and A. Subiantoro, "Net zero energy buildings in New Zealand: Challenges and potentials reviewed against legislative, climatic, technological, and economic factors," *J. Build. Eng.*, vol. 44, p. 102970, Dec. 2021, doi: 10.1016/J.JOBE.2021.102970.

[9] G. Holdmann, D. Pride, G. Poelzer, B. Noble, and C. Walker, "Critical pathways to renewable energy transitions in remote Alaska communities: A comparative analysis," *Energy Res. Soc. Sci.*, vol. 91, p. 102712, Sep. 2022, doi: 10.1016/J.ERSS.2022.102712.

[10] G. J. Bhagavathiammal, M. Lal, and K. Emperumal, "Observational evidence of equatorial ionospheric response to severe cyclonic storms 'AILA' and 'WARD' observed over the North Indian Ocean," *J. Atmos. Solar-Terrestrial Phys.*, vol. 211, p. 105462, Dec. 2020, doi: 10.1016/J.JASTP.2020.105462.

[11] P. Palomares-Morón, D. Vásquez, A. Escudero-Atehortua, and M. Toro, "Distributed generation: An inventory of installed capacity in Colombia through 2019," *2019 FISE-IEEE/CIGRE Conf. - Living Energy Transition, FISE/CIGRE 2019*, Dec. 2019, doi: 10.1109/FISECIGRE48012.2019.8985021.

[12] J. G. Rueda-Bayona, A. Guzmán, J. J. C. Eras, R. Silva-Casarín, E. Bastidas-Arteaga, and J. Horrillo-Caraballo, "Renewables energies in Colombia and the opportunity for the offshore wind technology," *J. Clean. Prod.*, vol. 220, pp. 529–543, May 2019, doi: 10.1016/J.JCLEPRO.2019.02.174.

[13] J. J. C. Eras *et al.*, "A look to the electricity generation from non-conventional renewable energy sources in Colombia," *Int. J. Energy Econ. Policy*, vol. 9, no. 1, pp. 15–25, 2019, doi: 10.32479/IJEEP.7108.

[14] A. Sagastume Gutiérrez, J. J. Cabello Eras, L. Hens, and C. Vandecasteele, "The energy potential of agriculture, agroindustrial, livestock, and slaughterhouse biomass wastes through direct combustion and anaerobic digestion. The case of Colombia," *J. Clean. Prod.*, vol. 269, p. 122317, Oct. 2020, doi: 10.1016/J.JCLEPRO.2020.122317.

[15] F. R. M. Arroyo and L. J. Miguel, "The Role of Renewable Energies for the Sustainable Energy Governance and Environmental Policies for the Mitigation of Climate Change in Ecuador," *Energies 2020, Vol. 13, Page 3883*, vol. 13, no. 15, p. 3883, Jul. 2020, doi: 10.3390/EN13153883.

[16] J. Barzola-Monteses, M. Mite-León, M. Espinoza-Andaluz, J. Gómez-Romero, and W. Fajardo, "Time Series Analysis for Predicting Hydroelectric Power Production: The Ecuador Case," *Sustain. 2019, Vol. 11, Page 6539*, vol. 11, no. 23, p. 6539, Nov. 2019, doi: 10.3390/SU11236539.

[17] W. E. Guastay Cajo and E. A. Llanes Cedeño, "THE USE OF HYDROPOWER FOR ELECTRIC POWER GENERATION AS A STRATEGY FOR INDUSTRIAL DEVELOPMENT IN ECUADOR," *Univ. Cienc. y Tecnol.*, vol. 24, no. 104, pp. 28–35, Sep. 2020, doi: 10.47460/UCT.V24I104.363.

[18] M. J. Pickl, "The renewable energy strategies of oil majors – From oil to energy?," *Energy Strateg. Rev.*, vol. 26, p. 100370, Nov. 2019, doi: 10.1016/J.ESR.2019.100370.

[19] M. Shadman *et al.*, "Ocean Renewable Energy Potential, Technology, and Deployments: A Case Study of Brazil," *Energies 2019, Vol. 12, Page 3658*, vol. 12, no. 19, p. 3658, Sep. 2019, doi: 10.3390/EN12193658.

[20] G. G. Dranka and P. Ferreira, "Towards a smart grid power system in Brazil: Challenges and opportunities," *Energy Policy*, vol. 136, p. 111033, Jan. 2020, doi: 10.1016/J.ENPOL.2019.111033.

[21] M. K. Bilonda, "Burning of biomass in the democratic republic of congo," *Biomass Burn. Sub-Saharan Africa Chem. Issues Action Outreach*, pp. 57–70, Jan. 2020, doi: 10.1007/978-940070808-2_5/FIGURES/4.

[22] R. Gad Mugagga, H. Baxter, and N. Chamdimba, "A Comprehensive Review on Status of Solar PV Growth in Uganda," *J. Energy Res. Rev.*, vol. 3, no. 4, pp. 1–14, Dec. 2019, doi: 10.9734/JENRR/2019/V3I430113.

[23] P. A. Trotter, N. J. Cooper, and P. R. Wilson, "A multi-criteria, long-term energy planning optimisation model with integrated on-grid and off-grid electrification – The case of Uganda," *Appl. Energy*, vol. 243, pp. 288–312, Jun. 2019, doi: 10.1016/J.APENERGY.2019.03.178.

[24] M. Takase, R. Kipkoech, and P. K. Essandoh, "A comprehensive review of energy scenario and sustainable energy in Kenya," *Fuel Commun.*, vol. 7, p. 100015, Jun. 2021, doi: 10.1016/J.JFUECO.2021.100015.

[25] M. Moner-Girona *et al.*, "Decentralized rural electrification in Kenya: Speeding up universal energy access," *Energy Sustain. Dev.*, vol. 52, pp. 128–146, Oct. 2019, doi: 10.1016/J.ESD.2019.07.009.

[26] A. George, S. Boxiong, M. Arowo, P. Ndolo, Chepsaigutt-Chebet, and J. Shimmon, "Review of solar energy development in Kenya: Opportunities and challenges," *Renew. Energy Focus*, vol. 29, pp. 123–140, Jun. 2019, doi: 10.1016/J.REF.2019.03.007.

[27] J. M. Hansen and G. A. Xydis, "Rural electrification in Kenya: a useful case for remote areas in sub-Saharan Africa," *Energy Effic.*, vol. 13, no. 2, pp. 257–272, Feb. 2020, doi: 10.1007/S12053018-9756-Z/FIGURES/10.

[28] C. Reilles, F. Tadeo, and M. Nachidi, "Powering an island resort by renewable energy - A feasibility analysis in the Maldives," *2019 8th Int. Conf. Syst. Control. ICSC 2019*, pp. 372–376, Oct. 2019, doi: 10.1109/ICSC47195.2019.8950599.

[29] M. Shumais, "Resource Use Efficiency of Electricity Sector in the Maldives," *J. Asian Financ. Econ. Bus.*, vol. 7, no. 1, pp. 111–121, Jan. 2020, doi: 10.13106/JAFEB.2020.VOL7.NO1.111.

[30] M. Khalil, M. A. Berawi, R. Heryanto, and A. Rizalie, "Waste to energy technology: The potential of sustainable biogas production from animal waste in Indonesia," *Renew. Sustain. Energy Rev.*, vol. 105, pp. 323–331, May 2019, doi: 10.1016/J.RSER.2019.02.011.

[31] M. T. Sambodo and R. Novandra, "The state of energy poverty in Indonesia and its impact on welfare," *Energy Policy*, vol. 132, pp. 113–121, Sep. 2019, doi: 10.1016/J.ENPOL.2019.05.029.
[32] A. Hidayatno, A. D. Setiawan, I. M. Wikananda Supartha, A. O. Moeis, I. Rahman, and E. Widiono, "Investigating policies on improving household rooftop photovoltaics adoption in Indonesia," *Renew. Energy*, vol. 156, pp. 731–742, Aug. 2020, doi: 10.1016/J.RENENE.2020.04.106.

[33] F. T. R. Silalahi, T. M. Simatupang, M. P. Siallagan, F. T. R. Silalahi, T. M. Simatupang, and M. P. Siallagan, "Biodiesel produced from palm oil in Indonesia: Current status and opportunities," *AIMS Energy 2020 181*, vol. 8, no. 1, pp. 81–101, 2020, doi: 10.3934/ENERGY.2020.1.81.

[34] D. Syafrianto, K. M. Banjar-Nahor, and N. Hariyanto, "Optimized design of renewable energy, firm capacity and stable operation power plant for serving off-grid commercial load in Indonesia," *EECCIS 2020 - 2020 10th Electr. Power, Electron. Commun. Control. Informatics Semin.*, pp. 21–26, Aug. 2020, doi: 10.1109/EECCIS49483.2020.9263446.

[35] M. G. Petterson and H. J. Kim, "Can Ocean Thermal Energy Conversion and Seawater Utilisation Assist Small Island Developing States? A Case Study of Kiribati, Pacific Islands Region," *Ocean Therm. Energy Convers. - Past, Present. Prog.*, May 2020, doi: 10.5772/INTECHOPEN.91945.

[36] B. Zalba, J. M. Marín, L. F. Cabeza, and H. Mehling, "Review on thermal energy storage with phase change: Materials, heat transfer analysis and applications," *Applied Thermal Engineering*, vol. 23, no. 3. Pergamon, pp. 251–283, Feb. 2003, doi: 10.1016/S1359-4311(02)00192-8.

[37] M. Viviana and O. L. Castillo, "Colombian energy planning - Neither for energy, nor for Colombia," *Energy Policy*, vol. 129, pp. 1132–1142, Jun. 2019, doi: 10.1016/J.ENPOL.2019.03.025.

[38] Unidad de Planeación Minero Energética – UPME, "PLAN ENERGÉTICO NACIONAL 2006-2025." Accessed: Mar. 08, 2022. [Online]. Available:

http://www.siel.gov.co/siel/Portals/0/PLAN_ENERGETICO_NACIONAL_2007.pdf.

[39] G. Carvajal-Romo, M. Valderrama-Mendoza, D. Rodríguez-Urrego, and L. Rodríguez-Urrego, "Assessment of solar and wind energy potential in La Guajira, Colombia: Current status, and future prospects," *Sustain. Energy Technol. Assessments*, vol. 36, p. 100531, Dec. 2019, doi: 10.1016/J.SETA.2019.100531.

[40] W. Weng, S. L. Becker, M. K. B. Lüdeke, and T. Lakes, "Landscape matters: Insights from the impact of mega-droughts on Colombia's energy transition," *Environ. Innov. Soc. Transitions*, vol. 36, pp. 1–16, Sep. 2020, doi: 10.1016/J.EIST.2020.04.003.

Modeling of a solar thermal system as a power supply alternative for a resistor water distillation system using TRNSYS

B. E. Tarazona-Romero[1 2], J. G. Ascanio-Villabona[1 2], A. D. Quintero-Ruiz[1 2], C. L. Sandoval-Rodriguez[1 3] and M. D. Duran-Sarmiento[2]

[1] Automation and Control Energy Systems Research Group (GISEAC), Faculty of Natural Sciences and Engineering, Electromechanical engineering, Unidades Tecnologicas de Santander (UTS), Student Street No 9-82, 680005, Bucaramanga, Colombia.

[2] Energy in Building research group (ENEDI), Doctoral Program in Energy Efficiency and Sustainability in Engineering and Architecture, Department of Machines and Thermal Engines, University of the Basque Country (UPV / EHU), Engineer Torres Quevedo Square, 1, 48013, Bilbao, BI, Spain.

[3] Doctoral Program in Electronic and Communications, Department of Communications Engineering, University of the Basque Country, UPV/EHU, 48013 Bilbao, Spain.

E-mail corresponding autor:btarazona@correo.uts.edu.co

Abstract. The sizing of a solar thermal system to feed the water distillers in the laboratory of the Santander Technological Units is presented, proposing a comparative study between three calculation methods (f-chart, instantaneous and ACSOL) for the estimation of the surface of solar capture, finally supported by modeling in the TRNSYS software of the final system, to evaluate its behavior dynamically during one year. Initially, a search for information is carried out to establish the models to develop each of the calculation methods, additionally technical data is collected from the laboratory equipment to determine the consumption of hot water. Subsequently, each of the calculation methods is applied in order to size the catchment surface, to finally carry out a comparative study between the results obtained, determining which is the most appropriate method for the calculation and defining the dimensions of the same, to develop a modeling of the dynamic behavior of the system through the TRNSYS Software. The final result presents a storage system with an average temperature of 62.13 ° C and solar collectors with an average temperature of 58.7 ° C for one year of operation. Finally, the operating time of the resistive stills is reduced from 11 hours a day to 6 hours with the integration of the Thermosolar system.

Keywords: Dynamic simulation, Solar Collectors, TRNSYS, Water Distillation

1. Introduction

The increase in energy demand in the countries grows as population increases [1] and technological developments [2], triggering high energy consumption, mostly from fossil fuels; being the biggest contributors to climate change [3]. For this reason, it is currently seeking to implement non-conventional renewable energy sources for the production of electrical or thermal energy, [4] [5], as a transitional alternative for the paradigm shift in the use of fossil fuels [6].

Thus, as non-conventional renewable energies [7] [8] will play an important role for humanity [9], being an alternative solution to supply in the future, percentages of energy demand worldwide. Eventually, various developments in the use of solar radiation through small-scale [10] or large-scale thermo-solar plants; integrated with photovoltaic [11] [12] or solar concentration [13] [14], systems, they have been implemented worldwide as a source of electrical and thermal energy production [15] [16].

Within the systems with high energy consumption, the production of hot water and / or steam at residential and industrial level is immersed, applied to processes such as; residential ACS systems, laundries, poultry sector, distillation, among others [17]. For their part, the distillation systems highly used in petrochemical processes, alcoholic beverages, the health sector, universities, among others, have a high energy consumption due to their generally resistive design or power using fossil fuels. Some authors [18], have developed direct solar distillation systems whose production capacity is on a small scale [19] and centralized systems applied to seawater distillation processes, for the production of fresh water, with high cost of implementation, operation and maintenance [20] [21].

Eventually, this work is aimed at the sizing of a solar exploitation system that allows the reduction of electrical energy consumption of the resistances used in the water distillation systems of the laboratories of the Santander Technology Units; that have a power of 3 kW each equipment. The design seeks to reduce the time of water treatment, which in normal working conditions oscillates in a production of 4 liters per hour, through the direct injection of the heat transfer liquid (Water) in the equipment, with an approximate inlet temperature of 60 ° C, in order to reduce the use and excessive consumption of it and taking advantage of the solar radiation of the City of Bucaramanga, Colombia.

The development of the system will be carried out through a comparative study of three different calculation methods that are: Indirect, F-chart and ASCOL 2.5, allowing a complete and detailed analysis of the system for a future implementation. Once the design of the system is in place, a modeling of its dynamic behavior will be carried out through the application of the TRNSYS software to know its operation for one year. The aim of this process is to contribute to the use of alternative and renewable generation means to reduce the high consumption of electrical energy by this equipment in percentage terms and, in turn, reduce greenhouse gas emissions, thereby reducing the environmental pollution generated in the facilities of the Santander Technology Units, through the future use of clean and inexhaustible energy from abundant solar radiation.

Finally, the document will be organized as follows. In section 2, the methodological aspects are presented, describing the analytical methods to be used to determine the size of the water heating system, as well as the TRNSYS modeling tool to evaluate the behavior for one year of operation of the system and its own characteristics. from the still. Section 3 covers the results of the system estimation and behavior modeling and finally section 4 presents the most relevant conclusions of the work.

2. Methods and Material

2.1. Analysis Methods

Currently there are various calculation methods for ACS systems, based on static, semi-static or empirical approximation procedures and dynamic simulation methods with a higher degree of complexity, which allow to determine the analysis in time intervals and controlled conditions of the system [22]. Next, three sizing methods for water heating systems are described, which are:

2.1.1. Indirect Method

Also known as the instantaneous method, it is based on using a calculation procedure in which all operating conditions are considered to be static. Being a manual procedure, this introduces a great error in the dimensioning, but in many cases, it serves as an initial starting point. Therefore, the energy demand is compared with the possible energy production that the installation would have if the environmental and operating conditions were kept constant throughout a period of time.

In this way, the following are presented for the determination of the surface of the ASC [23]:

- Determination of energy demand by applying the required% of solar coverage.
- Determination of the collector performance.
- Once the yield is known, obtain the energy production per unit area

IOP Conf. Series: Materials Science and Engineering 1299 (2024) 012011 doi:10.1088/1757-899X/1299/1/012011

- Once the energy production per unit area is obtained, we will divide the demand by the production to obtain the necessary area to produce said demand.

Finally, depending on the opening surface of each collector, determine the necessary number of solar collectors

2.1.2. *F-Chart Analysis*

The F-Chart method is an analysis that is useful for designing both forced circulation and thermosyphonic solar thermal systems, especially to select the size and type of solar collectors to supply that will meet the demands of DHW or other thermal loads. The objective of the method is to obtain the curve f and estimate the performance of a solar collector. Curve f represents the fraction of monthly heat load transformed from solar energy, being defined as a function of two parameters: absorbed energy (gain) and loss due to reflection in the solar collector [24].

The equation used in this method can be seen in the following formula [24].

$$f = 1{,}029\ D1 - 0{,}065\ D2 - 0{,}245\ (D1)^2 + 0{,}0018\ (D2)^2 + 0{,}0215\ (D1)^2 \qquad (1)$$

The sequence that is usually followed in the calculation is as follows:

- Assessment of the calorific loads for heating water destined for the production of DHW. or heating.
- Assessment of incident solar radiation on the inclined surface of the collector or collectors.
- Calculation of parameter D1
- Calculation of parameter D2
- Determination of the graph f.
- Assessment of monthly solar coverage.
- Assessment of the annual solar coverage and formation of tables

It is important to mention that for the calculation of the parameter f, which is the monthly energy fraction provided by the solar collection system, it must be above 60% for each month (or annual average); highlighting that no month can exceed 110% taking into account the technical code HE 4 of Spain, as a guide for the application of the current method.

2.1.3. *ACSOL 2.5*

The ACSOL calculation tool is the well-known free access TRNSYS simulation software from the University of Wisconsin-Madison [28], ACSOL contains a detailed mathematical model for each principal scheme and a database of meteorological conditions, accumulators, consumption profiles, etc. Through input menus the user provides additional information about the system to be studied; collection area, type of collector, volumes, consumptions, use profiles, configuration of obstacles, control method, etc. Once the case to be simulated has been defined, the computer solves the mathematical model and generates a results report.

2.2. *Modeling TRNSYS*

TRNSYS is a simulation software applied to the dimensioning and evaluation of renewable energy systems, which allows to analyze the operational behavior of the systems in a given time interval, under a series of variables that simulate environmental behaviors to which the devices are subjected in real environments.

Figure 1 presents the model of the ACS system to feed the resistive water distillation system.

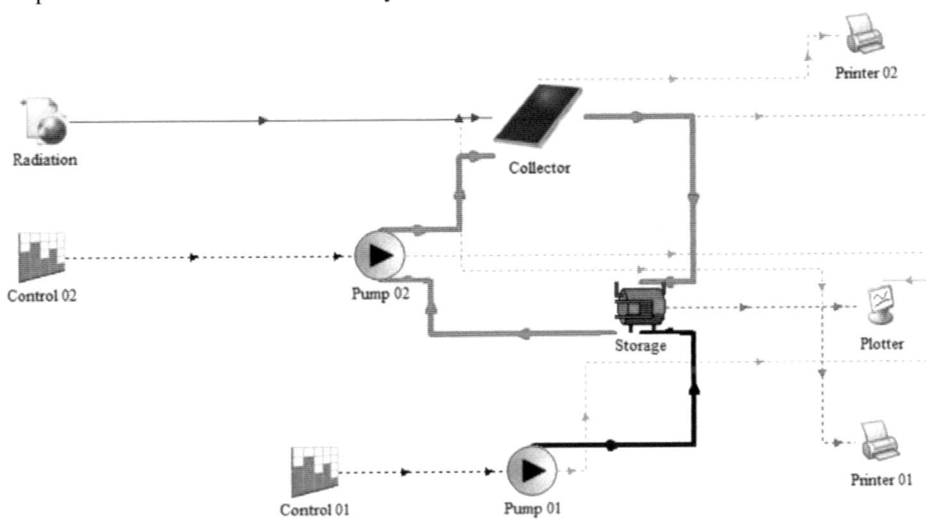

Figure 1. Modeling in TRNSYS

The following software types were used for the simulation process:

- Weather file: Radiation (Type), allows to link data from external weather files in TM2 format. To obtain these data, the Meteonorm software was used, geographic location, Bucaramanga, Colombia.
- Collector: Solar flat plate collector (Type), is used for heating water through the capture of solar radiation.
- Storage: storage and heat exchange system to finally supply the water to the resistive water distillation system.

2.3. *ACS consumption equipment*

The laboratories of the UTS have 3 distillers with the same technical characteristics described in Table 1, which reached a temperature of 97 ° C to carry out the distillation process. Currently, the consumption of distilled water by laboratories is approximately 15 gallons per day.

Table 1. Ficha Técnica Destilador BOECO 4L. Elaboración propia a partir del catálogo BOECO [25]

BOECO DEST 4L	
Distillate Flow	Approximate4l/h
Powe	3kw
Heating Element	Silica heater
Glass part material	Borosilicate glass3.3
Water consumption	Aprox 60l/hr
Weight	Aprox 4.5KG

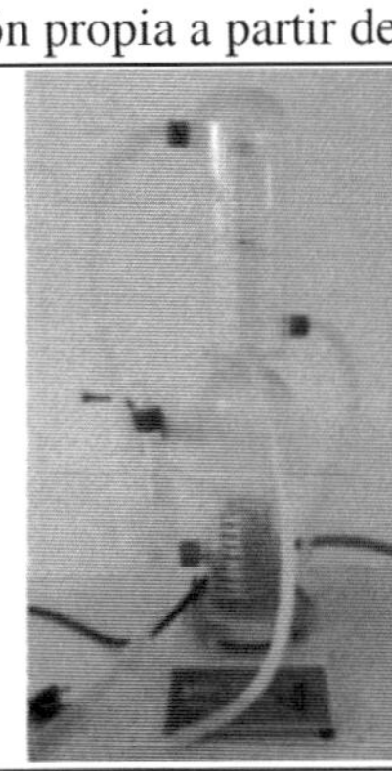

2.4. Solar Collector
A flat TS25 solar collector is selected with a yield of 63.236%, due to the support offered by the Termocol company, its efficiency, availability, cost and ease of online purchase.

2.5. Conditions for analytical methods
In order to apply the three methods following the same parameters, it was necessary to identify those variables that directly intervene in the efficiency of the design. Consequently, the table presents the conditions used in each of the three analytical methods, in order to size the initial ACS system and compare the results of the three estimates and finally determine the appropriate size.

Table 2. Condition for Sizing

DACS daily = Cac ·ρ·Cp·(Tac − Taf)	553080 KJ/day
Heat Carrier Liquid	Water 100%
Cac = Water density Hot water consumption (l / day)	3300
ρ = Water density (1 kg/l)	1
Cp = Specific heat of water (4.190 j/kg °C)	4190
Tac = Hot water temperature (°C)	60
Taf = Cold water temperature (°C)	20
Hot water supply time (hours / day)	6
Flow	60 L/h

Additionally, the system must have the following characteristics:
- Solar collection system
- Pumping system
- Storage tank
- Auxiliary systems

Finally, the following measures will be disregarded:
- Shadow losses
- Leaks through pipes

3. Results
Table 3 presents the main characteristics of the estimates of the ACS systems feeding the resistive distillation system of the laboratories of the Santander technological units, applying the analytical methods and the conditions described in section 2.5, highlighting:
- The instantaneous method projects a system composed of 19 solar collectors with an annual yield of 51.95% for a time of 6.67 hours of operation during the day. The amount of energy produced is 3,372 kWh / m ^ 2.
- The f-chart curves method was initially applied with the 19 collectors, based on the results obtained in the instantaneous method. The result was that no month of the year fulfilled the 60% coverage required by the application of the method to cover the demand for water from the distillation system. The result of the system performance was 19%.
- Subsequently, the method of f-chart curves based on an experimental test method was applied again in order to identify the number of collectors necessary to cover the minimum demand of HE 4, determining 81 collectors to cover with 61% of the annual yield. .

IOP Conf. Series: Materials Science and Engineering 1299 (2024) 012011 doi:10.1088/1757-899X/1299/1/012011

- Finally, the ACSOL 2.5 software was used where the conditions presented in Table 3 and the catchment surface determined by the instant no method were simulated. As a result, the system with 19 collectors presents a performance of 97.55% and Figure 2 presents the relevant data of the energy balance of the system.

Table 3. Anaitic Way

Analysis methods	Catchment Area $m2$	Collectors Quantity	% Annual coverage
Direct Method	45,6	19	51.95
F-chart curves	45,6	19	19
	194,6	81	61
ACSOL 2.5	45,6	19	97,55

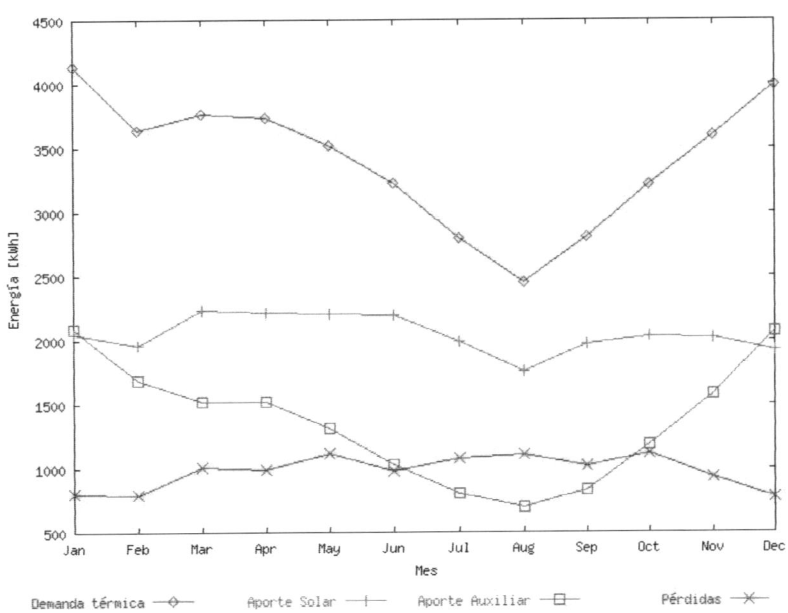

Figure 2. ACSOL 2.5 energy balance

For its part, Figure 3 presents the schematization of the system to be implemented for the supply of hot water to the resistive distillation system. Based on the results obtained by the three analytical methods (See Table 3), the system will be made up of 19 sensors and the connection circuit presented below.

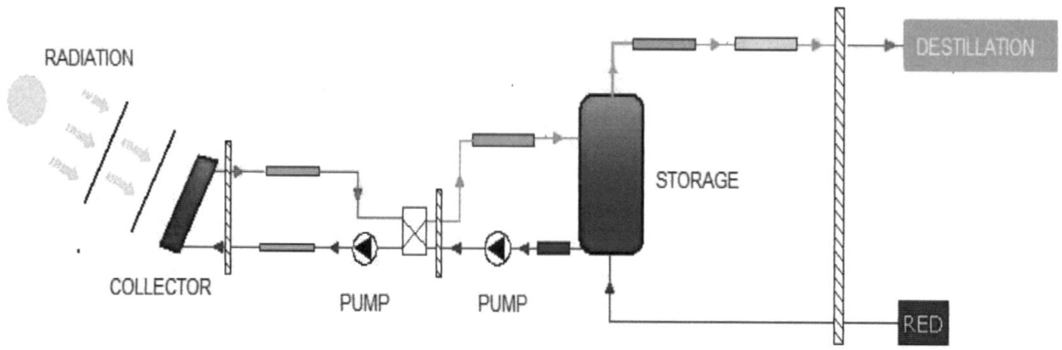

Figure 3. ACS system schematization

Once the size of the system was determined, the modeling was carried out in TRNSYS in order to look at the behavior of the system. Figure 4 presents the radiation data used in the simulation process and highlights:

- The average daily radiation of the city of Bucaramanga, Colombia is in a range of 4-8 kWh/m^2, allowing the collectors to reach temperatures of up to180°C.

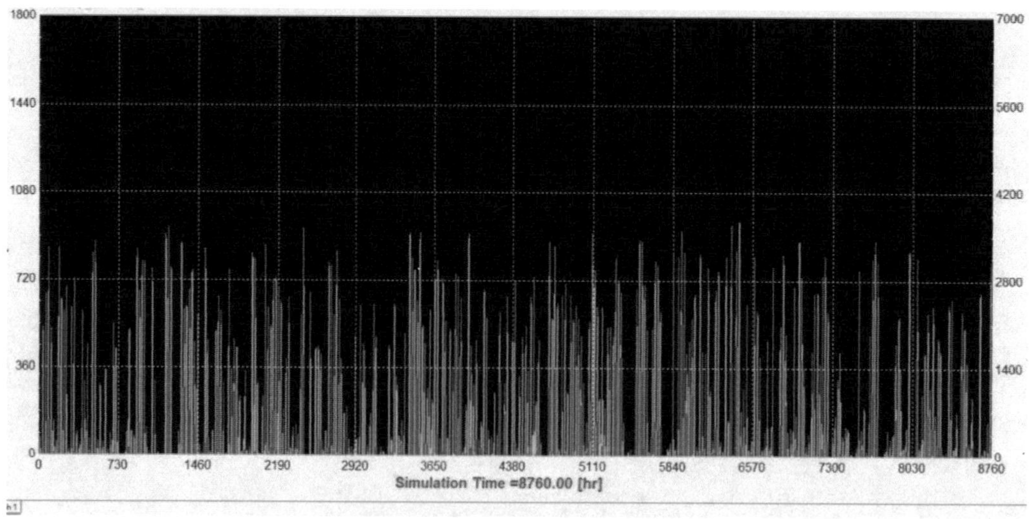

Figure 4. Solar Radiation Value

Figure 6 presents the behavior of the 19 connected solar collectors highlighting:

- The solar collectors will have a mixture of water vapor in some moments with temperatures of up to180°C.
- The average annual temperature of the liquid that enters the solar collection system and leaves it for the storage tank is58.7 °C.

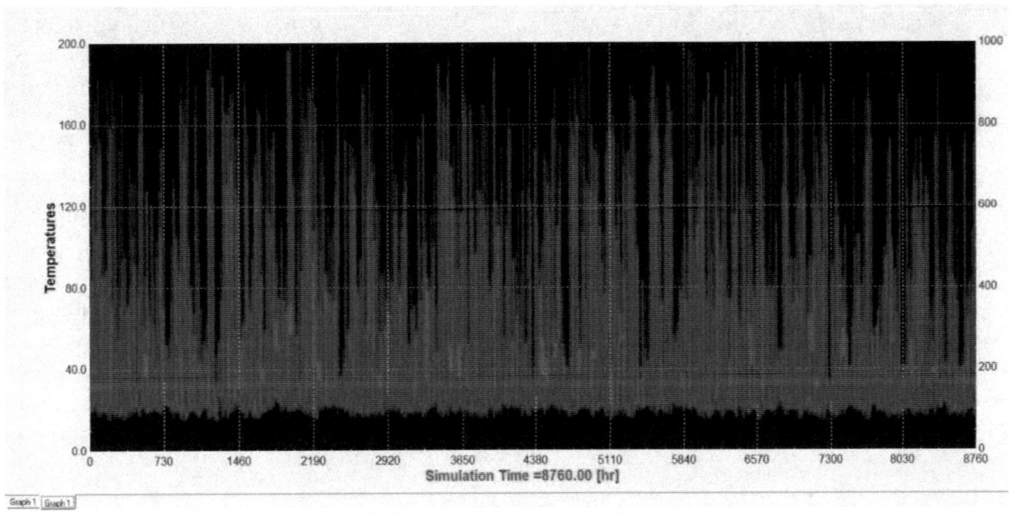

Figure 5. Solar Collector Outlet Temperature

Finally, Figure 6 presents the behavior of the storage system that would be connected to the resistive distiller of the Laboratories of the Technological Units of Santander, highlighting:

- The tank will store a mixture of water vapor in some moments with temperatures up to 120°C.
- The average annual temperature of the liquid that enters the tank and leaves the resistive still is 62.13 °C.

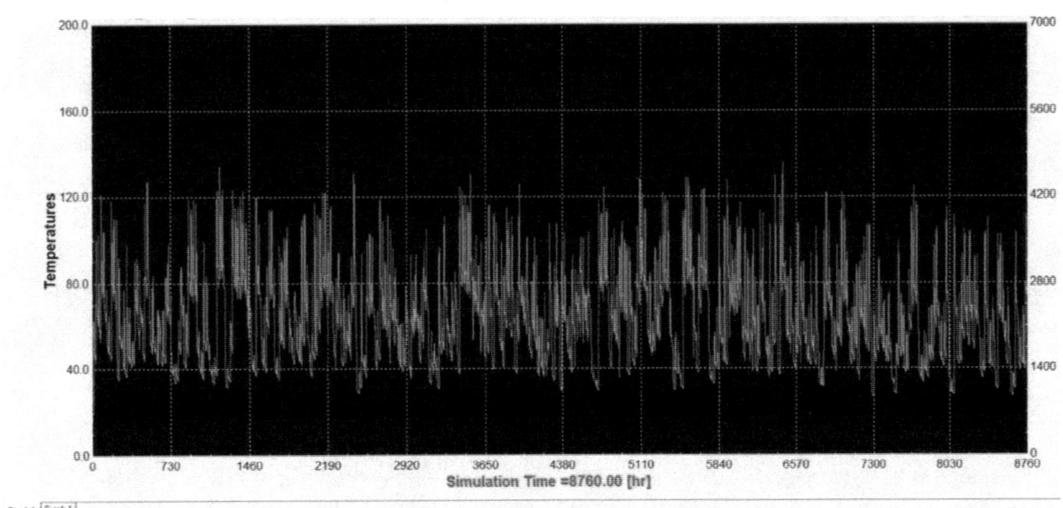

Figure 6. Storage outlet Temperature

4. Conclusions

Resistive distillation systems have high electrical energy consumption, which is why non-conventional renewable energies would be an alternative to supply the electrical and / or thermal needs of the devices. Currently, the drawback lies in the cost of implementing systems for high flow production of distilled water through technologies that take advantage of non-conventional renewable sources. This analysis seeks to provide an alternative to the high dependence on electricity supply for the production of distilled water and the most relevant conclusions are presented below:

- The analytical methods for the sizing of the ACS system allowed estimating the area of collectors required to supply the demand of the distillation system. The instantaneous method allowed to have an initial estimate with a coverage percentage of 51.95%. The method of the fchar curves was uncertain and its estimation differs from the other methods used. For its part, ACSOL 2.5 allowed to analyse the results obtained by the instantaneous method, resulting in a coverage of 97.55% and the final schematization of the system to be implemented.
- The modeling in TRNSYS allowed to dynamically evaluate the behavior of the dimensioned system through the analysis methods in the city of Bucaramanga, Colombia. The collection surface was 45.6 m ^ 2, equivalent to 19 collectors with an average annual temperature of 58.7 ° C at the outlet of the solar collectors and an average annual temperature in the storage tanks of 62.13 ° C.

Finally, the system dimensioned and subsequently modelled in TRNSYS is an alternative to be integrated into the resistive distillation system of the laboratories of the Santander Technological Units, satisfying the operating needs of the system and reducing the daily use of the still 11 hours. to 6 hours.

References

[1] J. Acquatella, «Energía y cambio climático: oportunidades para una política energética integrada en América Latina y el Caribe», dic. 2008, Accedido: 25 de agosto de 2020. [En línea]. Disponible en: https://repositorio.cepal.org//handle/11362/3635

[2] J. Morelos Gómez, «Análisis de la variación de la eficiencia en la producción de biocombustibles en América Latina», *Estudios Gerenciales*, vol. 32, n.º 139, pp. 120-126, abr. 2016, doi: 10.1016/j.estger.2016.01.001.

[3] M. A. Ise, S. C. Carrizo, y M. Forget, «Chapter 8 - Challenges of South American energy transition: energy efficiency and distributed generation», en *The Regulation and Policy of Latin American Energy Transitions*, L. Noura Guimarães, Ed. Elsevier, 2020, pp. 133-151. doi: 10.1016/B978-0-12-819521-5.00008-5.

[4] B. Sørensen, «3 - Individual renewable energy sources», en *Renewable Energy (Fifth Edition)*, B. Sørensen, Ed. Boston: Academic Press, 2017, pp. 219-353. doi: 10.1016/B978-0-12-8045671.00003-7.

[5] L. Noura Guimarães, «Chapter 1 - Is there a Latin American electricity transition? A snapshot of intraregional differences», en *The Regulation and Policy of Latin American Energy Transitions*, L. Noura Guimarães, Ed. Elsevier, 2020, pp. 3-20. doi: 10.1016/B978-0-12-819521-5.00001-2.

[6] P. Bajpai, «Chapter 3 - Renewable energy versus fossil resources», en *Biotechnology in the Chemical Industry*, P. Bajpai, Ed. Elsevier, 2020, pp. 23-30. doi: 10.1016/B978-0-12-8184028.00003-3.

[7] R. Vakulchuk, I. Overland, y D. Scholten, «Renewable energy and geopolitics: A review», *Renewable and Sustainable Energy Reviews*, vol. 122, p. 109547, abr. 2020, doi: 10.1016/j.rser.2019.109547.

[8] A. Aktaş y Y. Kirçiçek, «Chapter 6 - Solar Thermal Systems and Thermal Storage», en *Solar Hybrid Systems*, A. Aktaş y Y. Kirçiçek, Eds. Academic Press, 2021, pp. 127-137. doi: 10.1016/B978-0-323-88499-0.00006-9.

[9] M. Shao, Z. Han, J. Sun, C. Xiao, S. Zhang, y Y. Zhao, «A review of multi-criteria decision making applications for renewable energy site selection», *Renewable Energy*, vol. 157, pp. 377403, sep. 2020, doi: 10.1016/j.renene.2020.04.137.

[10] B. E. Tarazona-Romero, Á. Campos-Celador, Y. A. Muñoz-Maldonado, C. L. SandovalRodríguez, y J. G. Ascanio-Villabona, «Prototype of lineal solar collector Fresnel»:, *Visión electrónica*, vol. 14, n.º 1, Art. n.º 1, ene. 2020, doi: 10.14483/22484728.16013.

[11] A. Elamim, B. Hartiti, A. Haibaoui, A. Lfakir, y P. Thevenin, «Comparative study of photovoltaic solar systems connected to the grid: Performance evaluation and economic analysis», *Energy Procedia*, vol. 159, pp. 333-339, feb. 2019, doi: 10.1016/j.egypro.2019.01.006.

[12] J. G. Ascanio-Villabona, C. L. Sandoval-Rodriguez, A. D. Rincón-Quintero, B. E. TarazonaRomero, y R. E. Paez-Castro, «Building a prototype for functional analysis of the energy potential of the water flow in pipe ½ "using microturbines applied to Unidades Tecnológicas de Santander», *IOP Conf. Ser.: Mater. Sci. Eng.*, vol. 844, p. 012056, jun. 2020, doi: 10.1088/1757899X/844/1/012056.

[13] B. E. Tarazona Romero, A. C. Celador, C. L. S. Rodriguez, J. G. A. Villabona, y A. D. R. Quintero, «Design and construction of a solar tracking system for Linear Fresnel Concentrator», *Periodicals of Engineering and Natural Sciences (PEN)*, vol. 9, n.º 4, Art. n.º 4, oct. 2021, doi: 10.21533/pen.v9i4.1988.

[14] Y. Han, Y. Sun, y J. Wu, «A low-cost and efficient solar/coal hybrid power generation mode: Integration of non-concentrating solar energy and air preheating process», *Energy*, vol. 235, p. 121367, nov. 2021, doi: 10.1016/j.energy.2021.121367.

[15] B. E. Tarazona Romero, Y. A. M. Maldonado, A. C. Celador, y O. L. Pérez, «Optical Performance Assessment of a Handmade Prototype of Linear Fresnel Concentrator», *Periodicals of Engineering and Natural Sciences (PEN)*, vol. 9, n.º 4, Art. n.º 4, oct. 2021, doi: 10.21533/pen.v9i4.1987.

[16] A. Aktaş y Y. Kirçiçek, «Chapter 1 - Solar System Characteristics, Advantages, and Disadvantages», en *Solar Hybrid Systems*, A. Aktaş y Y. Kirçiçek, Eds. Academic Press, 2021, pp. 1-24. doi: 10.1016/B978-0-323-88499-0.00001-X.

[17] L. Betancur, J. Ascanio Villabona, y B. Tarazona Romero, *Diseño, construcción y validación teórico-experimental de un intercambiador de calor utilizando termosifones bifásicos*. 2021. doi: 10.13140/RG.2.2.15646.25926.

[18] G. Li, Q. Xuan, M. W. Akram, Y. Golizadeh Akhlaghi, H. Liu, y S. Shittu, «Building integrated solar concentrating systems: A review», *Applied Energy*, vol. 260, p. 114288, feb. 2020, doi: 10.1016/j.apenergy.2019.114288.

[19] B. E. Tarazona-Romero, A.-J. R. Nieves, C. L. Sandoval-Rodriguez, J. G. Ascanio-Villabona, M.A. Duran-Sarmiento, y A.-D. Rincon-Quintero, «Sizing of a sustainable condensation system for the production of sweet water through the use of photovoltaic panels in La Mesa de los Santos, Santander, Colombia», *IOP Conf. Ser.: Mater. Sci. Eng.*, vol. 1154, n.º 1, p. 012040, jun. 2021, doi: 10.1088/1757-899X/1154/1/012040.

[20] B. E. Tarazona-Romero, A. Campos-Celador, y Y. A. Maldonado-Muñoz, «Can solar desalination be small and beautiful? A critical review of existing technology under the appropriate technology paradigm», *Energy Research & Social Science*, vol. 88, p. 102510, jun. 2022, doi: 10.1016/j.erss.2022.102510.

[21] K. Liang, H. Zhang, H. Chen, D. Gao, y Y. Liu, «Design and test of an annular fresnel solar concentrator to obtain a high-concentration solar energy flux», *Energy*, vol. 214, p. 118947, ene. 2021, doi: 10.1016/j.energy.2020.118947.

[22] C. B. Barrera Gomez y E. Rodriguez Gozales, «Energias renovables Estudio de viabilidad de una instalacion solar fotovoltaica conectado a red para autoconsumo en la isla de Tenerife.», Pregrado, Universidad de la Laguna, 2016. Accedido: 26 de agosto de 2020. [En línea]. Disponible en: https://riull.ull.es/xmlui/bitstream/handle/915/2840/Energias%20renovables%20Estudio%20de%20viabilidad%20de%20una%20instalacion%20solar%20fotovoltaica%20conectado%20a%20red%20para%20autoconsumo%20en%20la%20isla%20de%20Tenerife.pdf;jsessionid=442AC76CC80E11EA5DC1CD7220F82C78?sequence=1

[23] A. Ruiz Gomez, «Energia Solar Termica: Aplicaciones, cálculo y diseño», en *Energia Solar*, España, 2010, p. 75.

IOP Conf. Series: Materials Science and Engineering 1299 (2024) 012011 doi:10.1088/1757-899X/1299/1/012011

[24] F. Obaco y J. Jaramillo, «Sistemas solar-térmicos: algoritmo operativo para aplicar el método F-Chart en la evaluación de colectores solares», p. 3, 2010.
[25] COTECNO, «Boeco Destilador 4 L » COTECNO | Equipamiento Cientifico | Prospecciones, Auscultation, Geofisica, Ingeniería», *COTECNO | Equipamiento Cientifico | Prospecciones, auscultation, geofisica, ingeniería*, 2020. https://www.cotecno.cl/nuestros-productos/boecodestilador-dest-4/ (accedido 26 de agosto de 2020).

Evaluation of technological tendencies in the carbon capture process: a review.

B. E. Tarazona-Romero[1 2], N. Y. Castillo-Leon[1], A. J. Rodriguez-Nieves, Y. P. Quizena-Fernandez, Y. C. Villabona-Niño [1]and O. Forero-Caceres [1],

[1] Automation and Control Energy Systems Research Group (GISEAC), Faculty of Natural Sciences and Engineering, Electromechanical engineering, Unidades Tecnologicas de Santander (UTS), Student Street No 9-82, 680005, Bucaramanga, Colombia.
[2] Energy in Building research group (ENEDI), Doctoral Program in Energy Efficiency and Sustainability in Engineering and Architecture, Department of Machines and Thermal Engines, University of the Basque Country (UPV / EHU), Engineer Torres Quevedo Square, 1, 48013, Bilbao, BI, Spain.

E-mail corresponding autor:btarazona@correo.uts.edu.co

Abstract. This paper evaluates the technological trends in the carbon sequestration process. For this purpose, these systems have been classified into two subsystems: conventional technologies subsystem and emerging technologies subsystem. Each is explored for its suitability for meeting a set of six attributes. A bibliometric analysis process was developed using the Scopus database and VOSviewer Software to present the potential of each subsystem evaluated, through an evaluation matrix. The analysis of the subsystems and attributes was performed through the formal concept analysis methodology (8FCA). To facilitate the processing of the information, the open access software concept Explorer was used. The analysis shows that conventional technologies, despite their cost, will be maintained and advance in their implementation process. The bibliometric analysis integrated with the applied FCA methodology has proven to be useful for the evaluation of technological typologies and serves as an alternative to develop theoretical studies that group and link different options as a model to evaluate a deterministic set of attributes.

Keywords: Bibliometric analysis, Carbon capture methods, Carbon capture technologies, Formal Concept Analysis.

1. Introduction

Industrial development and population growth have generated an increase in world energy demand, directly affecting global warming. Consequently, there has been an interest on the part of various world communities in progressively controlling anthropogenic CO2 emissions[1]. Renewable energies and carbon capture technologies are the two alternatives with the greatest short, medium and long term projection to contribute to the reduction of greenhouse gases[2]. Currently, carbon capture and storage presents commercial technological developments and research projects, which show the reliability of the process and the potential, with application to various industrial and residential sectors [3].

IOP Conf. Series: Materials Science and Engineering 1299 (2024) 012012 doi:10.1088/1757-899X/1299/1/012012

The Intergovernmental Panel on Climate Change (IPCC) will publish its Sixth Assessment Report (AR6) in late 2021-2022, which provides the most accurate estimate of temperature change over the next century, i.e., the data sets contributed approximately $0.1^{\circ}C$ for AR6 and AR5 [4]. To avoid a significant temperature increase, the IPCC estimates that global greenhouse gas emissions would need to be reduced by 50 to 80 percent by 2050.

Carbon dioxide emissions originate both naturally and from human activity. Biological processes such as decomposition, ocean release and respiration contribute to carbon dioxide emissions. The sources of air pollution from human activities are major industries such as cement production, electricity generation, food production, etc.[5]. Currently, renewable technologies are still under development. Although there are many solar, wind and hydro plants in operation, 80% of the world's energy is generated by fossil fuels. Carbon capture is a promising solution to reduce greenhouse gases as long as we rely on combustion-based power generation [6].

The term Carbon Capture and Storage (CCS) refers to the capture and storage of carbon dioxide (CO2) in stable subway reservoirs [7]. As a result of carbon sequestration, enhanced oil recovery (EOR) processes are also possible, involving the injection of CO2 into depleted oil fields to provide the pressure needed to recover more oil. The future availability of energy from fossil fuels, such as natural gas, is likely to be influenced by how effectively the associated CO2 emissions can be mitigated through carbon capture and sequestration (CCS). In turn, understanding how CCS affects the efficient recovery of energy from fossil fuel reserves in different parts of the world requires data on how the performance of each part of a particular CCS scheme is affected by technology-specific parameters and location-specific parameters, such as ambient temperature. This paper presents a study on how the energy consumption of an essential element of all CCS schemes, the CO2 compression process, varies with compressor design, CO2 pipeline pressure and cooling temperature.

Successful application of CCS technology will require overcoming many obstacles, including the need for government and industry to continue to advance the technology[8]. Another obstacle to the successful application of CCS technology is public opposition to the technology. Politicians, industry, and other stakeholders need to understand how the public perceives CCS, along with the factors that influence public acceptance or resistance to the technology[9].

For the theoretical development of this document, carbon capture technologies will be evaluated by applying a bibliometric analysis by means of specialized software and the information obtained will be evaluated by means of the method of formal analysis of concepts[10] [11] and the use of concept Explorer software [12], relating attributes and objects, simplifying information processing. With this in mind, the objectives of this work are:

i. Define the attributes that each carbon sequestration technology satisfies.
ii. Develop a review of the academic literature through databases and specialized software to evaluate the objects under the established attributes.
iii. Discuss the relationship between the technology and the established attributes through the use of the formal Concept Analysis.

This approach, the document is organized as follows. Section 2 presents the methodology, defining the bibliometric search string, as well as the attributes and assigning each technology an object for the development of the formal concept analysis. Section 3 covers the literature review on carbon capture and storage technologies, focusing on the level of compliance with each of the attributes. Section 4 presents the critical review using the Concept Explorer tool and the formal concept analysis method. Finally, Section 5 presents the most relevant conclusions of the paper

2. Methodology

La Carbon sequestration is an alternative to combat climate change. Commercially available technologies are available on the market; however, the high cost of operation and maintenance often generate an inefficient factor for their implementation. . For this reason, this work takes into consideration variables that consider relevant aspects for the implementation of commercial or emerging carbon capture technologies.

Initially, a specialized bibliometric analysis is carried out to recognize the articles with the highest incidence in the subject of study to feed the FCA methodology. The database used was Scopus and the program used to manage the information was the VOSviewer software as shown in Figure 1.

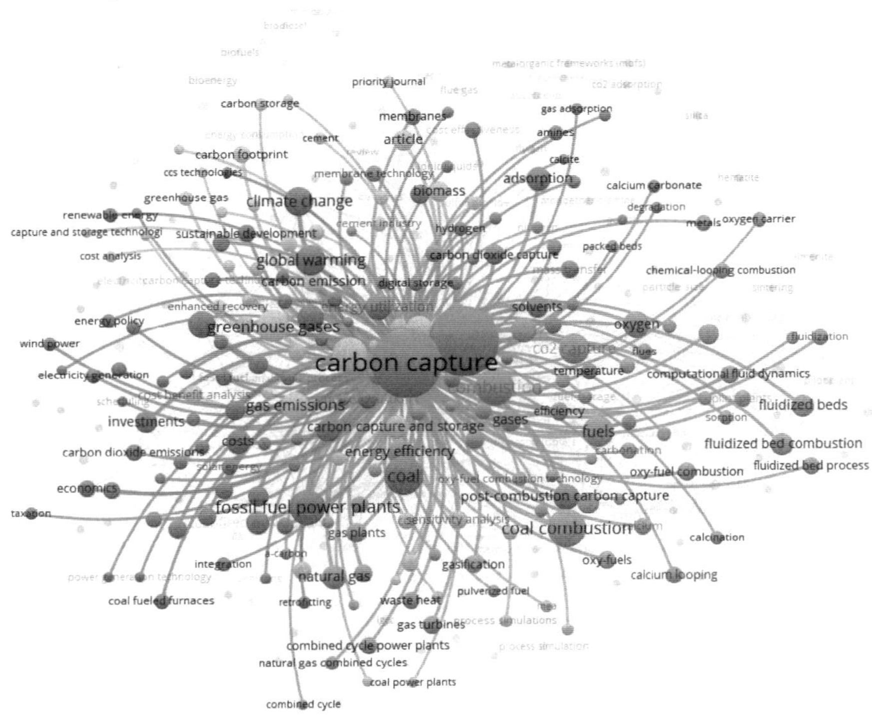

Figure 1. Bibliometric Analysis

For its part, the formal analysis of concepts is based on the fulfillment to a greater or lesser extent of a series of attributes with respect to a series of objects [13]. The following are the attributes that each carbon sequestration technology system must meet:

A1. Conventional technology
A2. Emerging technology
A3. Post-combustion
A4. Pre-combustion
A5. Oxy-combustion
A6.Technological Maturity at Industrial Level

Considering two technological categories: conventional and emerging, a literature review is carried out considering the attributes. Specifically, each of the characteristics that meet each attribute to a greater or lesser extent. This will result in a matrix that presents the degree to which each object or solar tracking system complies with the attributes, which will allow a qualitative analysis.

In this work, the evaluation of the information is carried out by applying the formal concept analysis (FCA) method, which allows the data to be analyzed by relating them conceptually[11]. The data analysis will be performed through the open access tool "Concept Explorer" [12], evaluating all existing relationships, grouping technological typologies and attributes.

3. Carbon Sequestration: Methods and Technologies

Combustion power plants produce 85% of the world's electricity [13]. Fuel is burned with air in a furnace or boiler, producing high-pressure gas or steam that drives the turbine generator [14]. The exhaust gases leaving these power plants consist mainly of nitrogen, CO_2 and water vapor. Some SO_2

and NOx are also produced, depending on the fuel type. CO_2 and other pollutants must be removed before the flue gas enters the atmosphere. In general, there are 3 methods to capture CO_2 [15]; post-combustion, pre-combustion and oxy-combustion. These methods produce a high density pure CO2 stream that can be stored or used for other industrial purposes.

Post-Combustion: This method is used on flue gas after combustion and before release to the atmosphere in fossil fuel-based power plants. This method is often implicit in industry due to its simplicity and flexibility [15] [16]. In a coal-fired power plant, the exhaust gases leaving the boiler are first treated to control air pollution and remove sulfur content. Then a chemical reaction (absorption) with a solvent such as monoethanolamine (MEA) is used to capture CO_2 [14] [17]. This process has two main pieces of equipment, namely absorber and stripper. About 90% of the CO_2 is captured in the absorber. The CO_2-loaded solvent then goes to the stripper, where heat is applied to release a high-concentration CO_2 stream. The CO_2 stream must be compressed for transport to the storage site. The MEA lean solvent is returned to the absorber[14] [18].

Pre-combustion: this method captures CO_2 prior to combustion. In a coal-fired power plant, coal first reacts with steam and oxygen at high temperature and pressure in a process called gasification, which produces synthetic gas or synthesis gas (CO and H_2)[19]. Then, the syngas is reacted with steam (H_2O) in a two-state change reactor that is converted to CO_2 and H_2. The CO_2 is captured by Selexol (a glycol-based solvent) using the physical absorption method. The H_2 is used for electricity production in a combined cycle as NGCC[14] [20]. Pre-combustion can also be used for natural gas power plants. Natural gas is converted to synthesis gas in a reaction with O_2 and H_2O called reforming. The synthesis gas then goes to a displacement reactor and produces CO_2 and H_2. This is the method used to manufacture H_2 [13] which can be used as a fuel for automobiles or as an energy source for industrial purposes.

Oxy-combustion: is an alternative to post-combustion for coal-fired power plants[21] [13]. In this method, coal is burned with pure oxygen instead of air, which eliminates nitrogen. The flue gases consist mainly of CO_2 and water vapor and a small amount of other pollutants such as SO2 and NOx. The water vapor is removed by the cooling and compression process, followed by the removal of other pollutants. The result is a pure CO_2 stream that is ready for storage. This method does not require an expensive post-combustion carbon capture system, but it does require an air separation unit (ASU) for oxygen production[13] [22].

CO_2 capture is a global problem, however, there is no simple generalized solution that can be applied worldwide. It is essential to analyze the origin of the CO_2 and which technology fits the specific needs before developing a project at any scale[23]. Currently, there are Conventional [24] and emerging technologies, which are adapted to various agricultural and industrial processes [25] [26].

Additionally, Table 1 presents the technologies applied in the carbon sequestration process and their relationship with each method.

Table 1. CO2 capture technologies

Type	**Method**	Technology
Conventional	Post-combustion	Absorption Adsorption Cryogenic Membrane
	Pre-combustion	Loop chemical combustion
	Oxi-combustión	
Emerging		Hydrate based separation Direct Air Mineralization of CO_2

Source: Own elaboration, information taken from [26] [27]

3.1. Conventional Technologies

3.1.1. Absorption (Ab). Absorption is the only well-used technology and the only commercial technology used in different industries due to costs. Absorption technology for CO_2 capture (ACC) has been around since the 1930s and is considered a mature technology. The technology has been used commercially to capture CO_2 from oil, coal, natural gas, thermal plants and industrial processes for more than five decades [28]. In the process of CO_2 capture by absorption, the separation of CO_2 from the flue gas is performed by a liquid sorbent through extraction or regeneration by a heating or depressurization process [29]. Figure 2 represents a basic technology implementation scheme for a Carbon Capture by absorption, according to Henrys' Law.

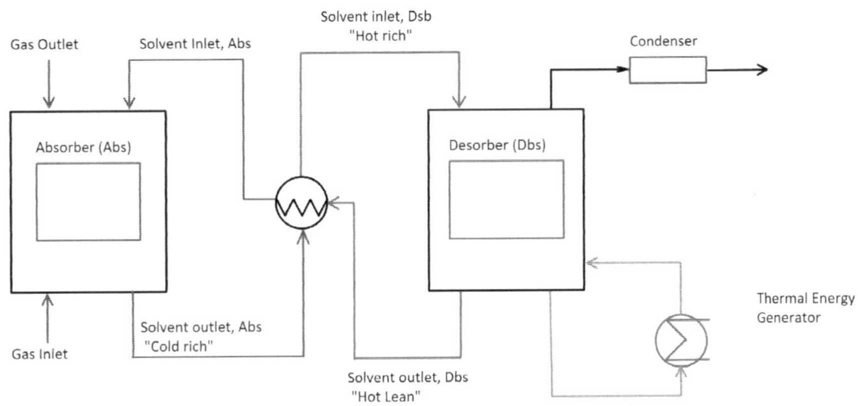

Figure 2 Carbon sequestration flowchart
Source: image taken from [30]

The absorbing liquid interacts with the gas and performs the absorption process in gas-liquid phase. Generally, the most widely applied sorbent is monoethanolamine (MEA) with an absorption efficiency of more than 90%[31] [32] [33]. However, there are other sorbents such as: diglycoamine (DGA), di-2-propanolamine (DIPA), methyl diethanolamine (MDEA), diethanolamine (DEA), triethanolamine (TEA), triethylenetetraamine (TETA), sodium carbonate (Na_2CO_3), ammonia (nH_3), potassium hydroxide (K_2CO_3), glucosamine (GEA) [26]. Eventually, modified ionic liquids were used for the CO_2 capture process as liquid adsorbent, presenting lower efficiencies than systems using MEA[26] [34].

The liquid solvents used for the absorption process have individual limitations and are not suitable for centralized systems. Consequently, some researchers emphasize that the mixture of two or more liquid solvents improves the energetic conditions and Absorption characteristics[35] [36] [37]. Recientemente, el solvente no acuoso Amina mejoró el consumo de energía del proceso de regeneración del solvente [38]. Recently, the non-aqueous solvent Amine improved the energy consumption of the solvent regeneration process [38]. On the other hand, Guo et al. carried out a novel experimental process, varying the traditional amine-water adsorbent by MEA-glycol, presenting reductions of up to 55% in the energy required by the adsorbent in the regeneration cycle [39]. This same year, Hwang et al, performed an experimental analysis, comparing the performance of a water-poor solvent K2Sol and MEA, highlighting that K2Sol requires 35% less energy than MEA for the regeneration process [40].

In general, two requirements should be considered when selecting the sorbent: (i) Reactivity of the gas and liquid, (ii) Solubility of the gas in the liquid. ACC technology has advantages such as [41] process reliability, high absorption efficiency, modularity when integrated into processes compared to other technologies [42], system without human operation [43], and purity levels exceeding 95% [41]. However, it still has limitations to be applied in power plants, as well as low process performance, high energy demand, and high investment, operation and maintenance costs [33], limiting its application at the commercial level [41].

Finally, it is of vital importance to highlight the need to continue improving the degradation process of the liquid amine solvent, due to the presence of HCL, HF, SO_2 and NO_2 in the composition of the combustion gases, incurring in: loss of liquid, generation of volatile gases, high energy demand, corrosion and low performance of the system, as well as the generation of Nitramines and Nitrosamines, affecting the environment and the health of living beings[44]. Kozak et al, presented the application of cooled ammonia that uses the ammonium base to separate CO_2, reducing the degradation of amines [45]. However, research studies and technological developments are still required to improve the current absorption process and the aforementioned problems.

Adsorption (Ad). It is considered one of the most promising technologies in the CO_2 capture process[46]. The process consists of separating CO_2 from flue gases by means of a solid adsorbent [47]. The adsorbent material must have a high surface area, high selectivity and high regeneration capacity. They generally include solid adsorbent amines based on carbon, graphite, zeolite and polymer [48]. The adsorbent material should have high surface area, high selectivity and high regeneration capacity. They generally include carbon-based solid adsorbent amines, organometallic structures, silica, solid ionic adsorbents, magnesium oxide, calcium oxide and LDH-derived mixed oxides [49].

Structurally, the absorption technology consists of a packed column filled with spherical absorber material. The flue gas flows through the column and CO_2 are attracted to the adsorbent material. The process can be performed in several ways [29]: (i) pressure swing adsorption (PSA), (ii) electrical swing adsorption (ESA), (iii) temperature swing adsorption (TSA), (iv) Hybrid Process (PTSA). On the other hand, the adsorption process is divided into two main categories: physisorption and chemisorption, i.e., the relationship of the physical or chemical bond of the gas to the adsorbent surface [50]. Finally, the process can be energetically profitable, but it depends on the proper selection of the adsorbent [51].

3.1.2. Membrane separation (MD). Membrane technology has a low removal efficiency and low CO_2 purity[52]. Its application is limited because it does not achieve a high degree of gas separation and the working volume is small [53]. Therefore, it requires multiple stages to carry out an adequate process. The drawbacks of membrane separation are [54]:

i. Membrane moisture: Moisture directly affects the efficiency of the process, it generates an increase of resistance in the mass transfer process.
ii. Solvent volatility: It affects the technology environmentally and economically. Therefore, the use of non-volatile solvents such as ionic liquids is required.
iii. Long-term stability: It directly affects the integrity of the membrane, due to the chemical reaction that may exist between the membrane material, the solvent and temperature changes.
iv. Dirty gas stream: the presence of other gas particles reduces the mass transfer ratio between the membrane and the CO_2.

Finally, further studies should be conducted to develop composite membranes [55], which are a combination of two or more materials. that generate a better ratio by removing gases, decreasing manufacturing and implementation costs [56], combination with other existing processes to increase the overall efficiency of the process [57].

3.1.3. Chemical Buckle Combustion (CLC). Chemical looping combustion (CLC) was first implemented in 1993 by Richter and Knoche to capture CO_2, under the idea of diluting CO_2 in the flue gas to avoid its direct interaction with air [58]. Generally, in a CLC system, the combustion air and fuel never interact with each other, creating a CO_2 exhaust stream that is not diluted with N_2, which facilitates CO_2 separation [59]

The process reaches temperatures between 800-1200 °C and pressures of 1-69 atm depending on the size of the CLC system[60]. In turn, it generally uses transition metal oxides such as Iron, Nickel, Cobalt, Copper, Calcium [29]. The main limitation of CLC technology is the variation of reactor pressure, incurring air leakage [61]. Additionally, there is a continuous interaction between the metal oxides with the unburned carbon, decreasing the yield [62]. However, the energy loss during CO_2 separation is low, CO_2 separation is easy and its application can be in liquid, solid and gaseous fuels[63].

Finally, it is important to highlight that CLC technology is currently under research and technological development, hoping to overcome the shortcomings of commercial plants and the need to desulfurize the fuel before exposing it to the reactor [60] [64].

3.2. Emerging Technologies

3.2.1. Hydrate-based gas separation. Hydrate-based gas separation (HBGS) is a relatively new and promising technology for capturing CO_2 from flue gases. The process exposes exhaust gases composed of CO_2 to water at high pressures, producing hidrates [65] [66].

The main disadvantage of this technology is the additional cost of equipment to treat large volumes of flue gas with low CO_2 content. In large-scale systems, energy consumption can sometimes be high, which reduces the profit margin [29]. However, the process has a high CO_2 storage capacity, energetically and environmentally sustainable, and is capable of capturing H_2S and CO simultaneously. [29].

3.2.2. Direct CO2 air mineralization It is a promising alternative for carbon capture. The process consists of accelerating the natural reaction of CO_2 from the air and minerals rich in magnesium or calcium in order to obtain a solidified and processable carbonate, to regenerate minerals, and be reused, in turn generating a stream of captured CO_2 [67]. The technology is under development and involves improvements in engineering processes to reduce the cost of permanent removal of high quality carbon and the resulting CO_2 could be stored for later application [67].

Finally, Table 2 presents the matrix of attributes for the analysis of the different carbon sequestration technologies. The attributes that meet each attribute are identified with an X in the matrix.

Table 1. Selection matrix of attributes with respect to objects

At / Obj	A1	A2	A3	A4	A5	A6
Ad	X		X			X
Ab	X		X			X
CP	X		X			
MD	X		X			
CLC	X			X		
HGBS		X			X	
DAM		X			X	

Source: Author

4. Results

Figure 3 represents the graphical diagram of the formal concept analysis performed in the concept Explorer tool for the analysis of carbon capture and storage technologies, based on the information contained in the attribute's matrix in Table 1. This diagram relates the small scale plants; classified in low and micro scale with those 6 attributes proposed to identify their advantages and disadvantages.

The different nodes shown in Figure 3 are identified as follows:

i. Blue-white: contains only attributes
ii. Blue-black: they are containers of solar tracking technologies related to attributes
iii. White: connector node

Considering the above, the analysis of the diagram is simple. The carbon capture and storage technologies analyzed are related to all the attributes that connect them from top to bottom, up to the source node. Specifically, Attribute A1 is directly related to CLC, MD, CP, Ad and Ab technologies. DAM and HGBS technologies do not meet attribute 1, and do not share characteristics with the other technologies. In general, Ad and Ab technologies have the highest number of attributes fulfilled (3).

From this analysis, it is evident that the two types of carbon capture and storage technologies with greater maturity, development and application at present are Adsorption and Absorption. There are other commercial technologies and applicable to processes, but they lack maturity at a centralized and commercial level, in some specific cases.

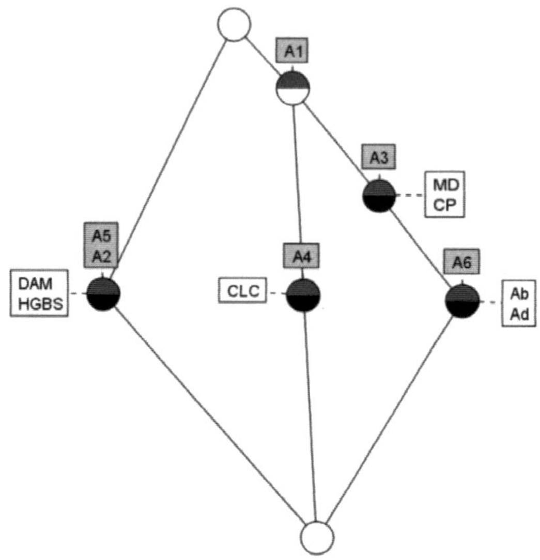

Figure 3. Formal Concept Analysis General Diagram
Source: own elaboration, image generated with concept Explorer software.

Figure 4 shows the FCA diagram that relates the carbon capture and storage technologies under a connecting line, i.e., under the compliance or relationship of attributes with each other or individual compliance. The Ad and Ab technologies lead in compliance with attributes A1, A3 and A6, while the MD and Cp technologies do not comply with attribute A6. However, both are commercial technologies at centralized and centralized level, only that Ad and Ab technologies have higher maturity at centralized implementations level.

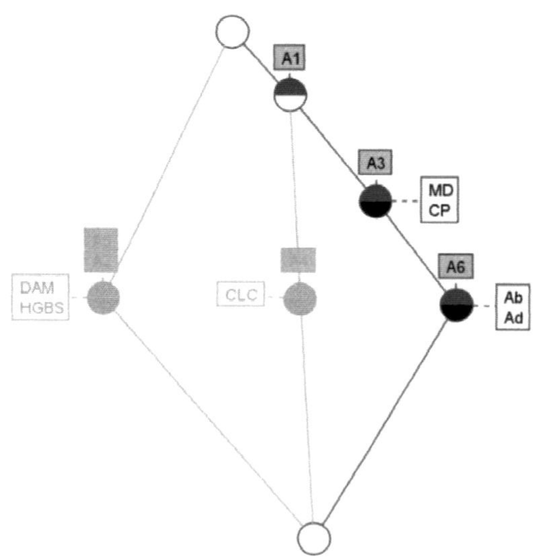

Figure 4. Diagram 01 Formal Analysis of Concepts
Source: own elaboration, image generated with concept Explorer software.

Figure 4 shows the FCA diagram that relates the carbon capture and storage technologies under a connecting line, i.e., under the compliance or relationship of attributes with each other or individual compliance. The Ad and Ab technologies lead in compliance with attributes A1, A3 and A6, while the MD and Cp technologies do not comply with attribute A6. However, both are commercial technologies at centralized and centralized level, only that Ad and Ab technologies have higher maturity at centralized implementations level.

On the other hand, Figure 5 presents the FCA diagram for CLC technology, which shares attribute A1 with MD, CP, Ad and Ab technologies; however, it is the only technology that complies with attribute A4. CLC is a commercial technology with low and medium scale applications.

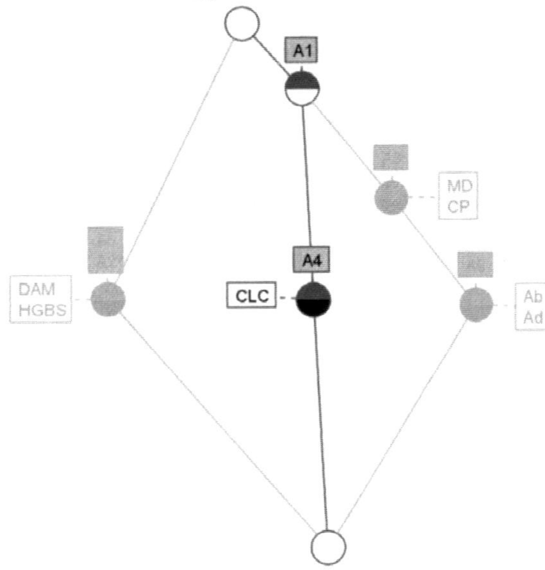

Figure 5. Diagram 02 Formal Analysis of Concepts
Source: own elaboration, image generated with concept Explorer software.

Finally, Figure 6 shows the FCA diagram for the DAM and HGBS technologies, highlighting the lack of sharing of any attribute with another technology. The DAM and HGBS technologies lack maturity at the commercial level; their developments are limited to experimentation in laboratories and low-scale systems under development.

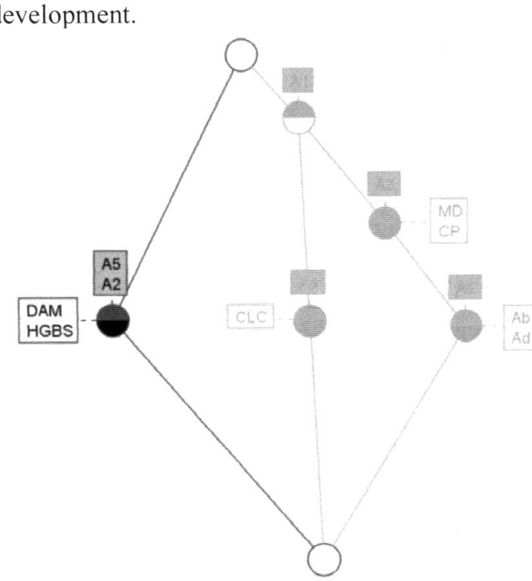

Figure 6. Diagram 03 Formal Analysis of Concepts
Source: own elaboration, image generated with concept Explorer software.

5. Conclusions

Carbon capture and storage technologies are currently presented as a promising technique to reduce and prevent greenhouse gas emissions at industrial and residential levels, through the implementation of various technologies at small, medium and large scales. This review summarizes the progress and technological changes. Relevant conclusions from the theoretical analysis are presented below.

i. It is important to highlight the efforts made by several researchers to evaluate the energy efficiency of sorbents based on the performance of the chemical reaction; however, it would be necessary to thermodynamically evaluate the consumption in order to obtain more conclusive data.

ii. The application of a single solvent generally shows a low absorption capacity for carbon sequestration, presenting shortcomings due to high levels of toxicity, flammability, decomposition, corrosivity, among others. However, the use of additives or mixtures with other solvents considerably improves its performance.

iii. It is important to mention that liquid and solid solvents are a highly efficient alternative in the absorption process. However, it is necessary to continue improving the degradation processes of liquids and solids in order to reduce the environmental impact generated by the process. However, the benefit they offer for the carbon capture process far outweighs the environmental drawbacks of solvents.

iv. Emerging technologies show potential in the medium and long term, but require significant efforts in the area of research and technological development, to solve the drawbacks derived from the few experimental results documented in an academic manner to date.

Finally, the academic and scientific information analyzed lends credibility to the quantitative analysis developed, although this type of analysis should not replace qualitative analysis, it is presented as an alternative to conventional studies.

References

[1] H. M. Sheikh, A. Ullah, K. Hong, y M. Zaman, «Thermo-economic analysis of integrated gasification combined cycle (IGCC) power plant with carbon capture», *Chemical Engineering and Processing - Process Intensification*, vol. 128, pp. 53-62, jun. 2018, doi: 10.1016/j.cep.2018.04.007.

[2] S. Fawzy, A. I. Osman, J. Doran, y D. W. Rooney, «Strategies for mitigation of climate change: a review», *Environ Chem Lett*, vol. 18, n.º 6, pp. 2069-2094, nov. 2020, doi: 10.1007/s10311-020-01059-w.

[3] F. Khosroabadi, A. Aslani, K. Bekhrad, y Z. Zolfaghari, «Analysis of Carbon Dioxide Capturing Technologies and their technology developments», *Cleaner Engineering and Technology*, vol. 5, p. 100279, dic. 2021, doi: 10.1016/j.clet.2021.100279.

[4] R. K. Singh *et al.*, «International Conference on Climate Risk Management, inputs for the Intergovernmental Panel on Climate Change's Sixth Assessment Report», *Climate and Development*, vol. 11, n.º 8, pp. 655-658, sep. 2019, doi: 10.1080/17565529.2018.1521331.

[5] S. R. Hari, C. P. Balaji, y K. Karunamurthy, «Carbon Capture and Storage Using Renewable Energy Sources: A Review», *IOP Conf. Ser.: Earth Environ. Sci.*, vol. 573, p. 012004, oct. 2020, doi: 10.1088/1755-1315/573/1/012004.

[6] S. Jackson y E. Brodal, «Optimization of the Energy Consumption of a Carbon Capture and Sequestration Related Carbon Dioxide Compression Processes», *Energies*, vol. 12, n.º 9, Art. n.º 9, ene. 2019, doi: 10.3390/en12091603.

[7] E. Parson y D. Keith, «Fossil Fuels Without CO2 Emissions», *Science*, vol. 282, n.º 5391, pp. 1053-1054, nov. 1998, doi: 10.1126/science.282.5391.1053.

[8] K. N. Finney, M. Akram, M. E. Diego, X. Yang, y M. Pourkashanian, «Chapter 2 - Carbon capture technologies», en *Bioenergy with Carbon Capture and Storage*, J. C. Magalhães Pires y A. L. D. Cunha Gonçalves, Eds. Academic Press, 2019, pp. 15-45. doi: 10.1016/B978-0-12-816229-3.00002-8.
[9] D. Pudasainee, V. Kurian, y R. Gupta, «2 - Coal: Past, Present, and Future Sustainable Use», en *Future Energy (Third Edition)*, T. M. Letcher, Ed. Elsevier, 2020, pp. 21-48. doi: 10.1016/B978-0-08-102886-5.00002-5.
[10] B. Ganter y R. Wille, *Análisis de concepto formal, Texto original: Fundamentos matemáticos*. Berlín Heidelberg: Springer-Verlag, 1999. Accedido: 9 de julio de 2021. [En línea]. Disponible en: https://www.springer.com/gp/book/9783540627715
[11] F. Sancho Caparrini, «Análisis Formal de Conceptos», *Análisis Formal de Conceptos*, 2017. http://www.cs.us.es/~fsancho/?e=78 (accedido 8 de julio de 2021).
[12] S. Yevtushenko, J. Tane, T. B. Kaiser, S. Objedkov, J. Hereth Correia, y H. Reppe, *Concept Explorer*. 2021. [En línea]. Disponible en: http://conexp.sourceforge.net/download.html
[13] T. Wilberforce, A. Baroutaji, B. Soudan, A. H. Al-Alami, y A. G. Olabi, «Outlook of carbon capture technology and challenges», *Science of The Total Environment*, vol. 657, pp. 56-72, mar. 2019, doi: 10.1016/j.scitotenv.2018.11.424.
[14] E. S. Rubin, H. Mantripragada, A. Marks, P. Versteeg, y J. Kitchin, «The outlook for improved carbon capture technology», *Progress in Energy and Combustion Science*, vol. 38, n.° 5, pp. 630-671, oct. 2012, doi: 10.1016/j.pecs.2012.03.003.
[15] K. H. M. Al-Hamed y I. Dincer, «A comparative review of potential ammonia-based carbon capture systems», *Journal of Environmental Management*, vol. 287, p. 112357, jun. 2021, doi: 10.1016/j.jenvman.2021.112357.
[16] L. Helei, P. Tantikhajorngosol, C. Chan, y P. Tontiwachwuthikul, «Technology development and applications of artificial intelligence for post-combustion carbon dioxide capture: Critical literature review and perspectives», *International Journal of Greenhouse Gas Control*, vol. 108, p. 103307, jun. 2021, doi: 10.1016/j.ijggc.2021.103307.
[17] M. Stec, A. Tatarczuk, T. Iluk, y M. Szul, «Reducing the energy efficiency design index for ships through a post-combustion carbon capture process», *International Journal of Greenhouse Gas Control*, vol. 108, p. 103333, jun. 2021, doi: 10.1016/j.ijggc.2021.103333.
[18] C. Ji, S. Yuan, M. Huffman, M. M. El-Halwagi, y Q. Wang, «Post-combustion carbon capture for tank to propeller via process modeling and simulation», *Journal of CO2 Utilization*, vol. 51, p. 101655, sep. 2021, doi: 10.1016/j.jcou.2021.101655.
[19] T. Zhou, H. Shi, X. Ding, y Y. Zhou, «Thermodynamic modeling and rational design of ionic liquids for pre-combustion carbon capture», *Chemical Engineering Science*, vol. 229, p. 116076, ene. 2021, doi: 10.1016/j.ces.2020.116076.
[20] A. G. Olabi *et al.*, «Assessment of the pre-combustion carbon capture contribution into sustainable development goals SDGs using novel indicators», *Renewable and Sustainable Energy Reviews*, vol. 153, p. 111710, ene. 2022, doi: 10.1016/j.rser.2021.111710.
[21] Y. Guan *et al.*, «Simulation study on the carbon capture system applying LNG cold energy to the O2/H2O oxy-fuel combustion», *Natural Gas Industry B*, vol. 5, n.° 3, pp. 270-275, jun. 2018, doi: 10.1016/j.ngib.2017.11.011.
[22] H. Sultan *et al.*, «Reducing the efficiency penalty of carbon dioxide capture and compression process in a natural gas combined cycle power plant by process modification and liquefied natural gas cold energy integration», *Energy Conversion and Management*, vol. 244, p. 114495, sep. 2021, doi: 10.1016/j.enconman.2021.114495.
[23] A. F. Ghoniem, «Needs, resources and climate change: Clean and efficient conversion technologies», *Progress in Energy and Combustion Science*, vol. 37, n.° 1, pp. 15-51, feb. 2011, doi: 10.1016/j.pecs.2010.02.006.
[24] M. Chawla *et al.*, «Membranes for CO2 /CH4 and CO2/N2 Gas Separation», *Chemical Engineering and Technology*, vol. 43, n.° 2, pp. 184-199, 2020, doi: 10.1002/ceat.201900375.

[25] P. Luis, T. Van Gerven, y B. Van der Bruggen, «Recent developments in membrane-based technologies for CO2 capture», *Progress in Energy and Combustion Science*, vol. 38, n.º 3, pp. 419-448, jun. 2012, doi: 10.1016/j.pecs.2012.01.004.

[26] A. Mukhtar *et al.*, «CO2 capturing, thermo-kinetic principles, synthesis and amine functionalization of covalent organic polymers for CO2 separation from natural gas: A review», *Journal of Natural Gas Science and Engineering*, vol. 77, p. 103203, may 2020, doi: 10.1016/j.jngse.2020.103203.

[27] S. Saqib *et al.*, «Influence of interfacial layer parameters on gas transport properties through modeling approach in MWCNTs based mixed matrix composite membranes», *Chemical Engineering Science*, vol. 218, p. 115543, jun. 2020, doi: 10.1016/j.ces.2020.115543.

[28] A. S. Bhown y B. C. Freeman, «Analysis and status of post-combustion carbon dioxide capture technologies», *Environmental Science and Technology*, vol. 45, n.º 20, pp. 8624-8632, 2011, doi: 10.1021/es104291d.

[29] A. Mukhtar *et al.*, «CO2 capturing, thermo-kinetic principles, synthesis and amine functionalization of covalent organic polymers for CO2 separation from natural gas: A review», *Journal of Natural Gas Science and Engineering*, vol. 77, p. 103203, may 2020, doi: 10.1016/j.jngse.2020.103203.

[30] K. H. M. Al-Hamed y I. Dincer, «A comparative review of potential ammonia-based carbon capture systems», *Journal of Environmental Management*, vol. 287, p. 112357, jun. 2021, doi: 10.1016/j.jenvman.2021.112357.

[31] A. L. Kohl y R. Nielsen, *Gas Purification*. Elsevier, 1997.

[32] S. Ma'mun, V. Y. Dindore, y H. F. Svendsen, «Kinetics of the reaction of carbon dioxide with aqueous solutions of 2-((2-aminoethyl)amino)ethanol», *Industrial and Engineering Chemistry Research*, vol. 46, n.º 2, pp. 385-394, 2007, doi: 10.1021/ie060383v.

[33] K. O. Yoro, M. K. Amosa, P. T. Sekoai, y M. O. Daramola, «Modelling and experimental investigation of effects of moisture and operating parameters during the adsorption of CO2 onto polyaspartamide», *International Journal of Coal Science and Technology*, vol. 6, n.º 2, pp. 225-234, 2019, doi: 10.1007/s40789-018-0224-3.

[34] F. Bougie y M. C. Iliuta, «CO2 absorption in aqueous piperazine solutions: Experimental study and modeling», *Journal of Chemical and Engineering Data*, vol. 56, n.º 4, pp. 1547-1554, 2011, doi: 10.1021/je1012247.

[35] S.-W. Park, J.-W. Lee, B.-S. Choi, y J.-W. Lee, «Absorption of carbon dioxide into non-aqueous solutions ofN-methyldiethanolamine», *Korean J. Chem. Eng.*, vol. 23, n.º 5, pp. 806-811, sep. 2006, doi: 10.1007/BF02705932.

[36] D.-H. Lee, W.-J. Choi, S.-J. Moon, S.-H. Ha, I.-G. Kim, y K.-J. Oh, «Characteristics of absorption and regeneration of carbon dioxide in aqueous 2-amino-2-methyl-1-propanol/ammonia solutions», *Korean Journal of Chemical Engineering*, vol. 25, n.º 2, pp. 279-284, 2008, doi: 10.1007/s11814-008-0049-7.

[37] S. A. Freeman, R. Dugas, D. H. Van Wagener, T. Nguyen, y G. T. Rochelle, «Carbon dioxide capture with concentrated, aqueous piperazine», *International Journal of Greenhouse Gas Control*, vol. 4, n.º 2, pp. 119-124, 2010, doi: 10.1016/j.ijggc.2009.10.008.

[38] J. Zhao, J. Fu, S. Deng, J. Wang, y Y. Xu, «Decoupled thermal-driven absorption-based CO2 capture into heat engine plus carbon pump: A new understanding with the case study», *Energy*, vol. 210, p. 118556, nov. 2020, doi: 10.1016/j.energy.2020.118556.

[39] H. Guo, C. Li, X. Shi, H. Li, y S. Shen, «Nonaqueous amine-based absorbents for energy efficient CO2 capture», *Applied Energy*, vol. 239, pp. 725-734, abr. 2019, doi: 10.1016/j.apenergy.2019.02.019.

[40] J. Hwang *et al.*, «An experimental based optimization of a novel water lean amine solvent for post combustion CO2 capture process», *Applied Energy*, vol. 248, pp. 174-184, ago. 2019, doi: 10.1016/j.apenergy.2019.04.135.

[41] K. O. Yoro, M. O. Daramola, P. T. Sekoai, E. K. Armah, y U. N. Wilson, «Advances and emerging techniques for energy recovery during absorptive CO2 capture: A review of process and non-process integration-based strategies», *Renewable and Sustainable Energy Reviews*, vol. 147, p. 111241, sep. 2021, doi: 10.1016/j.rser.2021.111241.

[42] M. Ozonoh, T. C. Aniokete, B. O. Oboirien, B. C. Udeh, K. O. Yoro, y M. O. Daramola, «Prediction of Emissions and Profits from a Biomass, Tyre, and Coal Fired Co-Gasification CHP Plant Using Artificial Neural Network: Nigerian and South African Perspectives», 2019, vol. 1378. doi: 10.1088/1742-6596/1378/2/022021.

[43] K. Goto, F. A. Chowdhury, H. Okabe, S. Shimizu, y Y. Fujioka, «Development of a low cost CO2 capture system with a novel absorbent under the COCS project», *Energy Procedia*, vol. 4, pp. 253-258, ene. 2011, doi: 10.1016/j.egypro.2011.01.049.

[44] H. Wang *et al.*, «Thermodynamic performance of solar-driven methanol steam reforming system for carbon capture and high-purity hydrogen production», *Applied Thermal Engineering*, vol. 209, p. 118280, jun. 2022, doi: 10.1016/j.applthermaleng.2022.118280.

[45] O. Kozák, K. K. R. Datta, M. Greplová, V. Ranc, J. Kašlík, y R. Zbořil, «Surfactant-Derived Amphiphilic Carbon Dots with Tunable Photoluminescence», *J. Phys. Chem. C*, vol. 117, n.º 47, pp. 24991-24996, nov. 2013, doi: 10.1021/jp4040166.

[46] X. Chen *et al.*, «Contrasting pathways of carbon sequestration in paddy and upland soils», *Global Change Biology*, vol. 27, n.º 11, pp. 2478-2490, 2021, doi: 10.1111/gcb.15595.

[47] Y. Zhang *et al.*, «Multi-walled carbon nanotubes/carbon foam nanocomposites derived from biomass for CO2 capture and supercapacitor applications», *Fuel*, vol. 305, p. 121622, dic. 2021, doi: 10.1016/j.fuel.2021.121622.

[48] D. Hospital-Benito, J. Lemus, R. Santiago, y J. Palomar, «Thermodynamic and kinetic evaluation of ionic liquids + tetraglyme mixtures on CO2 capture», *Journal of CO2 Utilization*, vol. 35, pp. 185-193, ene. 2020, doi: 10.1016/j.jcou.2019.09.015.

[49] H. Xi, M. Zhu, K. Y. Lee, y X. Wu, «Multi-timescale and control-perceptive scheduling approach for flexible operation of power plant-carbon capture system», *Fuel*, vol. 331, p. 125695, ene. 2023, doi: 10.1016/j.fuel.2022.125695.

[50] C. Even, D. Hadroug, Y. Boumlaik, y G. Simon, «Microalgae-based Bioenergy with Carbon Capture and Storage quantified as a Negative Emissions Technology», *Energy Nexus*, vol. 7, p. 100117, sep. 2022, doi: 10.1016/j.nexus.2022.100117.

[51] Z. Zhang, J. Du, K. Zhu, J. Guo, M. Li, y T. Xu, «Optimization scheduling of virtual power plant with carbon capture and waste incineration considering P2G coordination», *Energy Reports*, vol. 8, pp. 7200-7218, nov. 2022, doi: 10.1016/j.egyr.2022.05.027.

[52] A. G. Fink *et al.*, «Electrolytic conversion of carbon capture solutions containing carbonic anhydrase», *Journal of Inorganic Biochemistry*, vol. 231, p. 111782, jun. 2022, doi: 10.1016/j.jinorgbio.2022.111782.

[53] P. Shah, W. Wang, J. Z. Yang, L. Kahlor, y J. Anderson, «Framing climate change mitigation technology: The impact of risk versus benefit messaging on support for carbon capture and storage», *International Journal of Greenhouse Gas Control*, vol. 119, p. 103737, sep. 2022, doi: 10.1016/j.ijggc.2022.103737.

[54] H. Sun *et al.*, «XAS/DRIFTS/MS spectroscopy for time-resolved operando study of integrated carbon capture and utilisation process», *Separation and Purification Technology*, vol. 298, p. 121622, oct. 2022, doi: 10.1016/j.seppur.2022.121622.

[55] K. Yagihara, H. Ohno, A. Guzman-Urbina, J. Ni, y Y. Fukushima, «Analyzing flue gas properties emitted from power and industrial sectors toward heat-integrated carbon capture», *Energy*, vol. 250, p. 123775, jul. 2022, doi: 10.1016/j.energy.2022.123775.

[56] M. M. F. Hasan, M. S. Zantye, y M.-K. Kazi, «Challenges and opportunities in carbon capture, utilization and storage: A process systems engineering perspective», *Computers & Chemical Engineering*, vol. 166, p. 107925, oct. 2022, doi: 10.1016/j.compchemeng.2022.107925.

[57] F. Yu, X. Chu, D. Sun, y X. Liu, «Low-carbon economic dispatch strategy for renewable integrated power system incorporating carbon capture and storage technology», *Energy Reports*, vol. 8, pp. 251-258, nov. 2022, doi: 10.1016/j.egyr.2022.05.196.
[58] H. J. Richterr y K. F. Knoche, «Reversibility of Combustion Processes», en *Efficiency and Costing*, vol. 235, AMERICAN CHEMICAL SOCIETY, 1983, pp. 71-85. doi: 10.1021/bk-1983-0235.ch003.
[59] A. Cuadrat, A. Abad, L. F. de Diego, F. García-Labiano, P. Gayán, y J. Adánez, «Prompt considerations on the design of Chemical-Looping Combustion of coal from experimental tests», *Fuel*, vol. 97, pp. 219-232, jul. 2012, doi: 10.1016/j.fuel.2012.01.050.
[60] B. Moghtaderi, «Review of the Recent Chemical Looping Process Developments for Novel Energy and Fuel Applications», *Energy Fuels*, vol. 26, n.º 1, pp. 15-40, ene. 2012, doi: 10.1021/ef201303d.
[61] X. Xiao, W. Qin, J. Wang, J. Li, y C. Dong, «Effect of surface Fe-S hybrid structure on the activity of the perfect and reduced α-Fe2O3(001) for chemical looping combustion», *Applied Surface Science*, vol. 440, pp. 29-34, may 2018, doi: 10.1016/j.apsusc.2017.12.252.
[62] D.-A. Chisalita y A.-M. Cormos, «Dynamic simulation of fluidized bed chemical looping combustion process with iron based oxygen carrier», *Fuel*, vol. 214, pp. 436-445, feb. 2018, doi: 10.1016/j.fuel.2017.11.025.
[63] S. Abuelgasim, W. Wang, y A. Abdalazeez, «A brief review for chemical looping combustion as a promising CO2 capture technology: Fundamentals and progress», *Science of The Total Environment*, vol. 764, p. 142892, abr. 2021, doi: 10.1016/j.scitotenv.2020.142892.
[64] H. A. Alalwan y A. H. Alminshid, «CO2 capturing methods: Chemical looping combustion (CLC) as a promising technique», *Science of The Total Environment*, vol. 788, p. 147850, sep. 2021, doi: 10.1016/j.scitotenv.2021.147850.
[65] M. K. Mondal, H. K. Balsora, y P. Varshney, «Progress and trends in CO2 capture/separation technologies: A review», *Energy*, vol. 46, n.º 1, pp. 431-441, oct. 2012, doi: 10.1016/j.energy.2012.08.006.
[66] S. D. Kenarsari *et al.*, «Review of recent advances in carbon dioxide separation and capture», *RSC Advances*, vol. 3, n.º 45, pp. 22739-22773, 2013, doi: 10.1039/c3ra43965h.
[67] D. Babson, «Heirloom Carbon, ARPA_project», *Heirloom Carbon*, 2021. https://arpa-e.energy.gov/technologies/projects/transformative-low-cost-approach-direct-air-mineralization-co2-repeated (accedido 20 de noviembre de 2021).

Theoretical review based on the technologies developed in solar desalination.

K T Jaimes-Quintero[1], J G Ascanio-Villabona[1], B E Tarazona-Romero[1], A D Rincon-Quintero[1], M A Duran-Sarmiento[1].

[1] Unidades Tecnológicas de Santander UTS, Bucaramanga Santander 680005, Colombia

Ktjaimes@uts.edu.co, Jgascanio@correo.uts.edu.co, Btarazona@correo.uts.edu.co, Aricon@correo.uts.edu.co and Aduran@correo.uts.edu.co

Abstract. Increasing global demand for water and energy has given rise to renewable energies that offer an environmentally friendly alternative. Solar desalination systems have become a very attractive topic, due to the fact that areas lacking fresh water have a wealth of solar energy, allowing to present a convenient, promising and viable solution, also obtaining drinking water for consumption in remote areas from water with high salinity, where it is considered to be a process free of CO_2 emissions when powered by solar radiation, being sustainable and environmentally friendly. The main objective of this study is to provide an extensive review of the different solar desalination systems, evaluating the indicators, factors and technologies involved in the process, through scientific sources such as: articles, academic publications, international congresses, indexed journals, and others.

Keywords: Desalination, renewable energies, solar energy, integrated systems, technologies.

1. Introduction

Today, the shortage of drinking water and the increasing demand for energy is the greatest challenge facing [1], [2], due to the continuous increase in population, industrial activities and the fact that a quarter of mankind suffers from inadequate freshwater use [3], [4]. The water emergency situation is very alarming, mainly in arid and remote areas, despite the fact that the planet is covered by 97% of water, however, fresh water comprises only 2.5% of the total amount, the rest being seas, glaciers, ice caps and groundwater [5]

Brackish water has salinity in the range of 10,000 ppm, where the possible limit for drinking water is (500 to 1,000 ppm), so, to obtain these acceptable ranges is achieved through water desalination [6], [7], However, desalination is an energy-intensive process that requires 10,000 tons of fossil fuels per year to produce 1,000 cubic meters of water per day [8]–[10].

Eltawil et al. [11], refers that desalination is a process based on the separation of dissolved salts from seawater to convert it into fresh water suitable for consumption, industrial use or irrigation. [12]. At present, there are several desalination technologies or systems that, although they have different characteristics, such as type of energy, design and productivity, all meet the same objective, which is to desalinate water [13]–[16]. On the other hand, natural and human contamination of water does not imply that the total availability of water is of good quality and fit for consumption. That is why water quality is different for each region and depending on the type of contaminants subsoil [17].

Therefore, the integration of renewable energies with desalination systems is the most promising solution to overcome this water deficiency and, at the same time, to acquire the unprecedented commitment on a global scale [18], [19], to innovate water technologies and management systems, while preserving the quality of current resources, so as to reduce water demand by increasing efficiency and generally increasing the availability of freshwater quantity [20]–[24].

IOP Conf. Series: Materials Science and Engineering 1299 (2024) 012013 doi:10.1088/1757-899X/1299/1/012013

This work explores the global potential of desalination technologies integrated with renewable energy sources, mainly those related to solar radiation, by means of a narrative review method, identifying, compiling and selecting the systems with the greatest current impact, taking into account technological advances, increased productivity, efficiency and cost reduction.

Countries with desalination plants

The countries with the largest number of desalination plants are: United States with 2174 plants covering 34% worldwide, where 72% are reverse osmosis (RO), Saudi Arabia with 32% and 2086 plants of which 65% belong to multi-stage flash distillation (MSF), Japan covers 22% with 1457 plants and 90% of these belong to RO and lastly Spain with 12% and 760 plants of which 90% correspond to RO [25], [26]. (See Figure 1).

Figure 1. Countries with the largest installed desalination plants

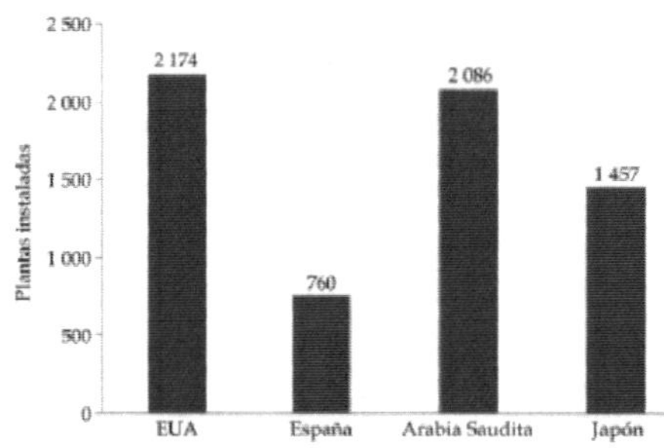

Source: [26]

2. Literature review

In order to initiate an efficient bibliographic search strategy, it was necessary to use a free and controlled language, so databases, mainly SCOPUS, were consulted, using the following search equation: "solar* AND desalination*" AND "technologies*" AND "review*". Searches are also carried out with the combination of the main keywords. The literature collected was extracted in .CSV format to perform the bibliographic analysis in the VOSviewer software. The figure 2 shows the co-occurrence of keywords in order to identify the research trend in the topic.

Figure 2. Keyword co-occurrence network

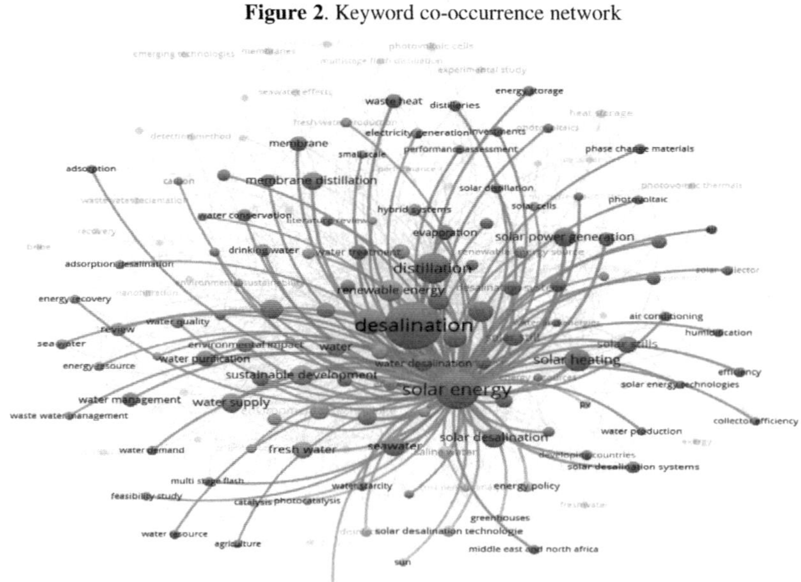

Source: Author, from VOSviewer software.

Solar desalination

In Mexico, according to Nassrullah et al. [22], there are 436 desalination plants installed in 320 locations, where only 5 use solar energy either as a power source or by using solar radiation as a heating source. Since solar desalination has become the best alternative in solving the problem of water scarcity and at the same time reducing carbon emissions while avoiding dangerous and irreversible uncontrollable effects on the world's climate and economy, sustainable energy systems, mainly solar radiation, play a very important role in the integration of renewable energy sources and efficient desalination technologies [22], there are several strategies that study this process, which are explained as follows:

2.1. Thermal System

In the thermal system, heat is needed to make the change from liquid to vapor, where the level of salinity of the water does not influence, i.e., seawater is heated until it evaporates, followed by vapor condensation, thus producing fresh water and leaving residues such as water with high salinity concentration [23]. The solar thermal system is classified into 2 processes, the direct process where there are the humidification-dehumidification (HDH) systems [24] and the solar stills system, on the other hand, the indirect process is composed by the multiple effect distillation (MED), multi-stage flash distillation (MSF) systems.

A. Humidification-dehumidification system (HDH)

A clear example of the HDH system is the production of rain, since the sun as a heat source is used to evaporate about 16 million tons of water from the earth into the atmosphere every second [27]. The vapor is carried by the wind to where it is treated through a zone and condenses at low temperature to form rain, snow, among others. The HDH method works in a similar way to the natura cycle (rain cycle), whereby the air is heated by a solar heater (SAH) in order to increase the vapor holding capacity where 1kg of air can carry 100g of water vapor at 60°C and rises to 500g at 80°C, whereby the air absorbs the salt water moisture in the humidifier and rejects it as fresh water in the dehumidifier [28], [29]. This HDH system consists of two main elements, humidifier and dehumidifier, which can be driven by a variety of heat sources (see Figure 3).

Figure 3. Schematic of simple HDH desalination system

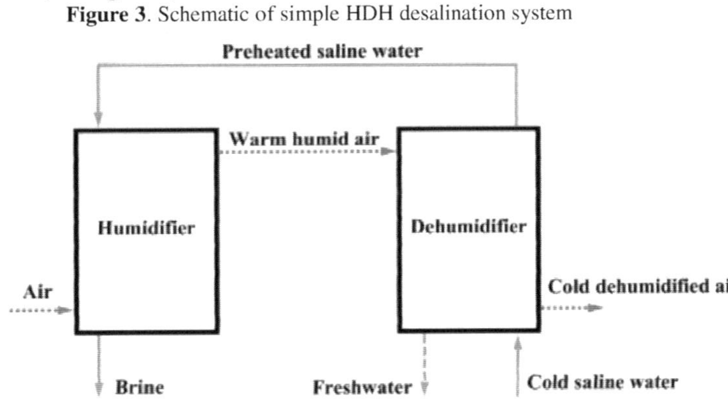

Source: [30]

Generally, there are different types of HDH configurations in order to improve the performance of each component in the system [30]–[34]. Hammadi [35], presents a theoretical study of the humidification - dehumidification (HDH) process in transient mode inside an open solar still, consisting of a longitudinal water pool, covered by a transparent glass panel with an inlet for dry air and an outlet for humid air. The results obtained show that freshwater productivity is highly affected by salt water, air currents and glass temperature.

The author Zhao et al. [36], presented a method using four-stage cross-flow dehydration and humidification (HDH) with direct contact dehumidifiers as shown in Figure 4, which results in a higher total water yield from the system as the air volume increases with air flow velocity; freshwater absorption has the highest yield of water per unit volume and the lowest cost of pure water. (See figure 4)

Over the years, studies have been conducted to improve the efficiency of the various desalination systems, however, problems of scale and cost have not achieved a great result, due to the fact that traditional components are built separately, which increases their manufacturing cost [36]–[39].

Figure 4. Schematic of the four-stage HDH desalination system with direct contact dehumidifiers

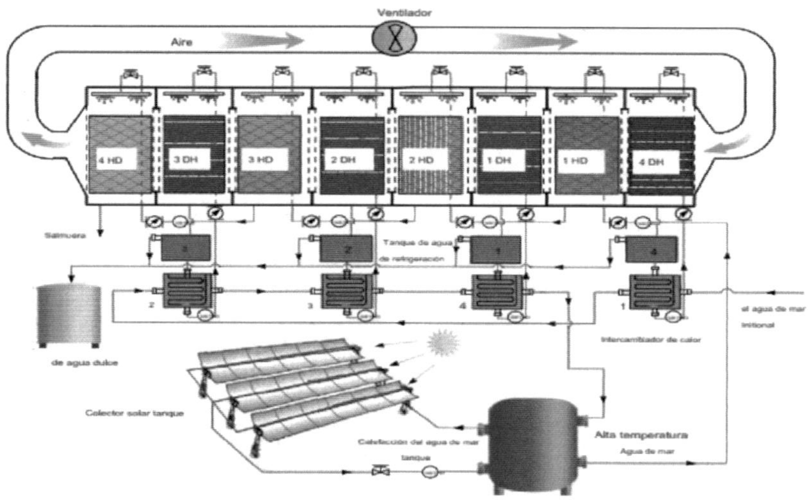

Source: [36]

B. Multi-effect System (MED)

MED plants consist of vertical or horizontal pipes, where the steam condenses on one side of the pipe while the other side releases the salt water caused by evaporation, this evaporated salt water is distributed over the outer pipes [40], [41]. In each effect of the MED process, heat exchange tubes are wetted with seawater, so that the steam flows through the tubes and condenses, thus producing fresh water.

2.2. Concentrated solar energy system (CSP)

Concentrated solar power (CSP) in the development of desalination has become the most viable and sustainable alternative as Mohammadi [42], since CSP generates heat at high temperatures, it can be used for electric power generation through power cycles or to satisfy heat needs for different applications such as desalination systems, both small and large scale for year-round fresh water production, mainly in regions of high solar radiation and saline water availability.

Research is currently focused on the search for alternative renewable energy sources. Concentrated solar power (CSP) is one of the most promising technologies for heat and power, due to its solar utilization [42], [43]

2.3. Solar collector

Solar collectors are based on optical principles of reflection in order to concentrate the solar radiation in a specific area called focus, so that the energy collected from the sun's radiation can be transferred to the water due to the thermal differential, causing it to change phase to clean it [17]. In these processes, the idea is to obtain better heat transfer results [44]– [47].

The solar desalination systems have been catalogued for their low efficiency due to heat loss on the collector surface, however, the parabolic trough collectors, since they concentrate the light in a single area, which makes the heat loss lower compared to ordinary collectors, have a higher efficiency [47]

Authors Mehrpooya, et al. [43], developed a concentrated solar power plant with desalination processes and with absorption cooling cycle designated to supply power, fresh water and cooling, aimed at reducing carbon and dioxide emissions. This system consists of a concentrated solar thermal plant with parabolic dish collectors and steam turbine, using a multi-effect desalination process with parallel seawater feed and a single-stage water and ammonia absorption cooling system, as shown in Figure 5, resulting in the destruction of 86% of the total energy rate of the system, with an efficiency of 66.07%, while the thermal efficiency of the system was 93.91%. That is, the weakness of the energy analysis to identify power quality and irreversibility's within the system. The economic results showed that the proposed integrated structure had a payback period of 5,738 years and a net annual gain of $6.828 million per year.

Figure 5. Schematic system of the integration of solar thermal plant, absorption cooling system and MED cycle.

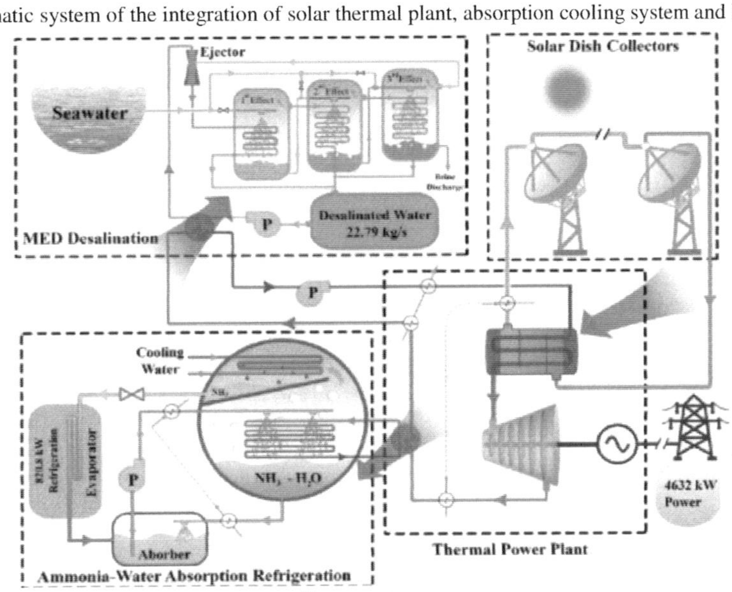

Source: [43]

2.4. Combined desalination and CCHP system (combined heat and cooling energy)

The integration of renewable energies with desalination and combined heating, cooling and power (CCHP) systems is one of the possible alternatives, using a two-tier methodology to optimize the system. The authors Luo, et.al. [48], assume a design on a remote island in the South China Sea, which consists of the displacement method, branch and bound method which are used to solve the optimization problem by mixed integer linear programming in the design and operation stages. As a result, the capacity of the electrical storage system is lower than the thermal capacity due to the high cost/performance ratio of the electrical storage, however, it can meet the energy demands, and the authors note that the integration of a water tank improves the productivity performance.

2.5. Solar chimneys

Solar chimneys are reliable devices with long operating life time, they are usually composed of a tower and an open collector on the edges, so, the incident radiation heats the ground under a collector, which in turn, heats the air inside the solar chimney, to generate a flow of hot air that can drive wind turbines and produce electricity, their maintenance costs are relatively low due to their robust and simple structure, the efficiency of these chimneys is very low, ranging from 0.5-10% of the solar energy delivery. However, there are not enough studies of the use of solar chimneys for desalination, there are only a few studies available, according to the authors Maia, et.al. [3], since, they focus on mathematical simulations and modeling, although the hybrid solar chimney-desalination system has great potential and can maximize performance, it is still a research phase.

Solar chimney plants have not had major demonstrations and have been produced intermittently with no improvement in performance, there are several projects internationally, for example, the 1500m high chimney system planned for South Africa, a 200 MW plant in China, three more 200 MW projects in the USA and a 200 MW solar chimney plant in China[49], these have only been planned and have not been built, although they promise to meet the necessary scales and better performance, taking into account that the stacks would have to have relatively low capital and investment delivery prices [50]

2.6. Combined system with renewable energy

Renewable energies are intended to provide a solution to the problems mentioned above, by integrating desalination systems with renewable energy sources, which are divided into two types:

- Desalination processes driven by heat from renewable energy systems.
- Membrane and distillation processes driven by electricity or mechanical power derived from renewable energies

Therefore, desalination systems integrated or hybrid with renewable energies are the most relevant issue today, allowing to face the scarcity of fresh water [21], [51], [52]. That is why, authors Kabeel & Said [17], offer a study on a modified solar-powered hybrid desalination system (SS-HDH). This consists of a solar still (SS) and the humidification-dehumidification (HDH) process, integrated with a solar water and air heater, where they studied the performance indicators and the total productivity of the system. Thus, this system has had a positive result on yield and productivity, because the maximum productivity was 18.25 l/m day with air and water mass flow rates. The humidification efficiency reached 79% and was affected by the increase in water mass flow rate rather than an increase in air mass flow rate, i.e., system performance is affected by air mass flow rate. Similarly, authors Morad et al. [53], conducted a literature search for methods or processes that improve the low performance of current desalination systems, where they found that increasing the salinity level of the water decreased productivity. Combined system with renewable energy.

2.7. Reverse osmosis (RO) desalination

The reverse osmosis (RO) process has several applications in industry, such as seawater desalination, food processing and wastewater treatment [15]. This separation method is the process with the highest efficiency and is environmentally friendly, while maintaining the same time with low energy consumption [22]. This RO process is based on the separation of two media with different solute concentrations through a pressure driven semi-permeable membrane, where water and some ion fractions can move through the membrane from the side with lower concentration to the side with higher concentration, while most salts are rejected.

Authors Wu, et.al. [23], developed a hybrid power system, based on reverse osmosis desalination coupled with a photovoltaic system and diesel fuel, in a power plant in order to meet load demand and increase water availability in Iran. Where they propose an optimal design of the autonomous photovoltaic hybrid system with diesel fuel, battery and reverse osmosis, using a metaheuristic technique based on tabu search, where results are obtained as a high efficiency being economical and environmentally friendly, however, being a single diesel system or a single photovoltaic system.

3. Conclusions

In this work, a theoretical review of solar desalination systems, their current technological status and the studies that focus on these systems was conducted; the review provides an overview of current desalination technologies and focuses on hybrid systems that take advantage of solar radiation, either for plants that produce only water or for freshwater cogeneration plants, the study presented that:

The CSP system can vary in temperature and produce electricity, which is suitable for driving a variety of desalination systems classified as either thermal or membrane processes, membrane processes are based on electrically guided concentration gradients such as RO, and thermal systems require evaporation of water to separate the salt, such as MED or MSF, in these thermally driven systems to improve performance and reduce costs it is feasible to use the TVC or an absorption heat pump.

These CSP-hybrid systems are classified according to their application for either cogeneration of energy or water-only production. The review showed that there are several possible ways to hybridize a

CSP, considering factors such as the appropriate selection of technologies (solar collectors), such as the study of the integrated desalination plant with a parabolic trough collector receiver tube, as well as showing high performance results in fresh water productivity, which was totally free of salts and suitable for consumption, improving its efficiency, due to its solar collector tracking system.

Thus, CSP-hybrid desalination plants are potentially cost-effective for the production of fresh water in an environmentally friendly manner, although significant research and advancement is still needed for these plants to be economically viable.

The studies presented various performance parameters for achieving high efficiency, such as overall thermal efficiency, freshwater productivity capacity, energy capacity, as well as evaluated the performance of various configurations and methods of operation to find the optimal scheme, considering the effect on performance as the location determines the level of available solar radiation and the characteristics of the brackish or seawater to be used, Therefore, these solar desalination systems (hybrid plants with solar energy), compared to traditional desalination plants powered by fossil fuels, show their great viability at present, contributing to reduce CO_2 pollution, and decreasing environmental problems due to depletion of natural resources.

In most cases, the lack of drinking water occurs in places with abundant solar radiation, which allows the use of various methods with solar energy, because desalination plants consume a high value of electricity, although desalination is not a polluting process, the generation of electricity to meet the demand makes it necessary to link the production of water with renewable energy sources.

Finally, reverse osmosis has an advantage over other thermal systems in that this process makes use of both brine and seawater, presenting relatively low investment costs depending on the water to be desalinated, although for this process the use of physicochemical treatment is indispensable, it can be coupled with another type of system and with renewable energies to supply the energy demand needed to operate the high-pressure pumps.

References

[1] N. R. Bond, R. M. Burrows, M. J. Kennard, and S. E. Bunn, "Water Scarcity as a Driver of Multiple Stressor Effects," *Multiple Stressors in River Ecosystems: Status, Impacts and Prospects for the Future*, pp. 111–129, Jan. 2019, doi: 10.1016/B978-0-12-811713-2.00006-6.

[2] C. Charcosset, "A review of membrane processes and renewable energies for desalination," *Desalination*, vol. 245, no. 1–3, pp. 214–231, Sep. 2009, doi: 10.1016/J.DESAL.2008.06.020.

[3] C. B. Maia, F. V. M. Silva, V. L. C. Oliveira, and L. L. Kazmerski, "An overview of the use of solar chimneys for desalination," *Solar Energy*, vol. 183, pp. 83–95, May 2019, doi: 10.1016/J.SOLENER.2019.03.007.

[4] C. M. and A. Yadav, "Water desalination system using solar heat: A review," *Renewable and Sustainable Energy Reviews*, vol. 67, pp. 1308–1330, Jan. 2017, doi: 10.1016/J.RSER.2016.08.058.

[5] W. Wang, S. Xie, Q. Pan, Y. Dai, R. Wang, and T. Ge, "Air-cooled adsorption-based device for harvesting water from island air," *Renewable and Sustainable Energy Reviews*, vol. 141, p. 110802, May 2021, doi: 10.1016/J.RSER.2021.110802.

[6] D. Curto, V. Franzitta, and A. Guercio, "A Review of the Water Desalination Technologies," *Applied Sciences 2021, Vol. 11, Page 670*, vol. 11, no. 2, p. 670, Jan. 2021, doi: 10.3390/APP11020670.

[7] E. Jones, M. Qadir, M. T. H. van Vliet, V. Smakhtin, and S. mu Kang, "The state of desalination and brine production: A global outlook," *Science of The Total Environment*, vol. 657, pp. 1343–1356, Mar. 2019, doi: 10.1016/J.SCITOTENV.2018.12.076.

[8] R. Santosh, H. S. Lee, and Y. D. Kim, "A comprehensive review on humidifiers and dehumidifiers in solar and low-grade waste heat powered humidification-dehumidification desalination systems," *J Clean Prod*, vol. 347, p. 131300, May 2022, doi: 10.1016/J.JCLEPRO.2022.131300.

[9] M. W. Shahzad, M. Burhan, L. Ang, and K. C. Ng, "Energy-water-environment nexus underpinning future desalination sustainability," *Desalination*, vol. 413, pp. 52–64, Jul. 2017, doi: 10.1016/J.DESAL.2017.03.009.

[10] H. Nassrullah, S. F. Anis, R. Hashaikeh, and N. Hilal, "Energy for desalination: A state-of-theart review," *Desalination*, vol. 491, p. 114569, Oct. 2020, doi: 10.1016/J.DESAL.2020.114569.

[11] M. A. Eltawil, Z. Zhengming, and L. Yuan, "A review of renewable energy technologies integrated with desalination systems," *Renewable and Sustainable Energy Reviews*, vol. 13, no. 9, pp. 2245–2262, Dec. 2009, doi: 10.1016/J.RSER.2009.06.011.

[12] B. Tomaszewska *et al.*, "Utilization of renewable energy sources in desalination of geothermal water for agriculture," *Desalination*, vol. 513, p. 115151, Oct. 2021, doi: 10.1016/J.DESAL.2021.115151.

[13] A. Ali, R. A. Tufa, F. Macedonio, E. Curcio, and E. Drioli, "Membrane technology in renewable-energy-driven desalination," *Renewable and Sustainable Energy Reviews*, vol. 81, pp. 1–21, Jan. 2018, doi: 10.1016/J.RSER.2017.07.047.

[14] A. Shafieian and M. Khiadani, "A novel solar-driven direct contact membrane-based water desalination system," *Energy Convers Manag*, vol. 199, p. 112055, Nov. 2019, doi: 10.1016/J.ENCONMAN.2019.112055.

[15] M. A. Al-Obaidi, J. P. Li, C. Kara-Zaïtri, and I. M. Mujtaba, "Optimisation of reverse osmosis based wastewater treatment system for the removal of chlorophenol using genetic algorithms," *Chemical Engineering Journal*, vol. 316, pp. 91–100, May 2017, doi: 10.1016/J.CEJ.2016.12.096.

[16] M. Al-Sahali and H. Ettouney, "Developments in thermal desalination processes: Design, energy, and costing aspects," *Desalination*, vol. 214, no. 1–3, pp. 227–240, Aug. 2017, doi: 10.1016/J.DESAL.2006.08.020.

[17] A. E. Kabeel and E. M. S. El-Said, "Experimental study on a modified solar power driven hybrid desalination system," *Desalination*, vol. 443, pp. 1–10, Oct. 2018, doi: 10.1016/J.DESAL.2018.05.017.

[18] S. Al-Amshawee, M. Y. B. M. Yunus, A. A. M. Azoddein, D. G. Hassell, I. H. Dakhil, and H. A. Hasan, "Electrodialysis desalination for water and wastewater: A review," *Chemical Engineering Journal*, vol. 380, p. 122231, Jan. 2020, doi: 10.1016/J.CEJ.2019.122231.

[19] D. Borge-Diez, F. J. García-Moya, P. Cabrera-Santana, and E. Rosales-Asensio, "Feasibility analysis of wind and solar powered desalination plants: An application to islands," *Science of The Total Environment*, vol. 764, p. 142878, Apr. 2021, doi: 10.1016/J.SCITOTENV.2020.142878.

[20] I. S. Al-Mutaz, "MSF challenges and survivals," *Desalination Water Treat*, vol. 177, pp. 14–22, 2019, doi: 10.5004/dwt.2020.24908.

[21] C. Ghenai, A. Merabet, T. Salameh, and E. C. Pigem, "Grid-tied and stand-alone hybrid solar power system for desalination plant," *Desalination*, vol. 435, pp. 172–180, Jun. 2018, doi: 10.1016/J.DESAL.2017.10.044.

[22] W. L. Ang, D. Nordin, A. W. Mohammad, A. Benamor, and N. Hilal, "Effect of membrane performance including fouling on cost optimization in brackish water desalination process," *Chemical Engineering Research and Design*, vol. 117, pp. 401–413, Jan. 2017, doi: 10.1016/J.CHERD.2016.10.041.

[23] B. Wu, A. Maleki, F. Pourfayaz, and M. A. Rosen, "Optimal design of stand-alone reverse osmosis desalination driven by a photovoltaic and diesel generator hybrid system," *Solar Energy*, vol. 163, pp. 91–103, Mar. 2018, doi: 10.1016/J.SOLENER.2018.01.016.

[24] M. Elimelech and W. A. Phillip, "The future of seawater desalination: Energy, technology, and the environment," *Science (1979)*, vol. 333, no. 6043, pp. 712–717, Aug. 2011, doi: 10.1126/SCIENCE.1200488/SUPPL_FILE/ELIMELECH.SOM.PDF.

[25] F. E. Ahmed, R. Hashaikeh, and N. Hilal, "Solar powered desalination – Technology, energy and future outlook," *Desalination*, vol. 453, pp. 54–76, Mar. 2019, doi: 10.1016/J.DESAL.2018.12.002.

[26] B. Moossa, P. Trivedi, H. Saleem, and S. J. Zaidi, "Desalination in the GCC countries- a review," *J Clean Prod*, vol. 357, p. 131717, Jul. 2022, doi: 10.1016/J.JCLEPRO.2022.131717.
[27] B. Gabra, M. Rady, A. M. Abdel Ghany, and Mohamed. A. Shamseldin, "Modelling and control of solar-driven humidification–dehumidification desalination plant," *Journal of Electrical Systems and Information Technology 2019 6:1*, vol. 6, no. 1, pp. 1–16, Dec. 2019, doi: 10.1186/S43067-019-0006-Z.
[28] S. M. Zubair, M. A. Antar, S. M. Elmutasim, and D. U. Lawal, "Performance evaluation of humidification-dehumidification (HDH) desalination systems with and without heat recovery options: An experimental and theoretical investigation," *Desalination*, vol. 436, pp. 161–175, Jun. 2018, doi: 10.1016/J.DESAL.2018.02.018.
[29] N. A. S. Elminshawy, F. R. Siddiqui, and M. F. Addas, "Experimental and analytical study on productivity augmentation of a novel solar humidification–dehumidification (HDH) system," *Desalination*, vol. 365, pp. 36–45, Jun. 2015, doi: 10.1016/J.DESAL.2015.02.019.
[30] M. Faegh, P. Behnam, and M. B. Shafii, "A review on recent advances in humidificationdehumidification (HDH) desalination systems integrated with refrigeration, power and desalination technologies," *Energy Convers Manag*, vol. 196, pp. 1002–1036, Sep. 2019, doi:
10.1016/J.ENCONMAN.2019.06.063.
[31] M. Mehrpooya, M. Mohammadi, and E. Ahmadi, "Techno-economic-environmental study of hybrid power supply system: A case study in Iran," *Sustainable Energy Technologies and Assessments*, vol. 25, pp. 1–10, Feb. 2018, doi: 10.1016/J.SETA.2017.10.007.
[32] P. Malek, J. M. Ortiz, and H. M. A. Schulte-Herbrüggen, "Decentralized desalination of brackish water using an electrodialysis system directly powered by wind energy," *Desalination*, vol. 377, pp. 54–64, Jan. 2016, doi: 10.1016/J.DESAL.2015.08.023.
[33] S. Liyanaarachchi, L. Shu, S. Muthukumaran, V. Jegatheesan, and K. Baskaran, "Problems in seawater industrial desalination processes and potential sustainable solutions: A review," *Rev Environ Sci Biotechnol*, vol. 13, no. 2, pp. 203–214, Nov. 2014, doi: 10.1007/S11157-0139326-Y/TABLES/5.
[34] D. U. Lawal, M. A. Antar, A. Khalifa, and S. M. Zubair, "Heat pump operated humidificationdehumidification desalination system with option of energy recovery," *https://doi.org/10.1080/01496395.2019.1706576*, vol. 55, no. 18, pp. 3467–3486, Dec. 2020, doi: 10.1080/01496395.2019.1706576.
[35] S. H. Hammadi, "Theoretical analysis of humidification – dehumidification process in an open type solar desalination system," *Case Studies in Thermal Engineering*, vol. 12, pp. 843–851, Sep. 2018, doi: 10.1016/J.CSITE.2018.09.009.
[36] Y. Zhao, H. Zheng, S. Liang, N. Zhang, and X. long Ma, "Experimental research on four-stage cross flow humidification dehumidification (HDH) solar desalination system with direct contact dehumidifiers," *Desalination*, vol. 467, pp. 147–157, Oct. 2019, doi: 10.1016/J.DESAL.2019.06.003.
[37] A. Mahmoud, H. Fath, and M. Ahmed, "Enhancing the performance of a solar driven hybrid solar still/humidification-dehumidification desalination system integrated with solar concentrator and photovoltaic panels," *Desalination*, vol. 430, pp. 165–179, Mar. 2018, doi: 10.1016/J.DESAL.2017.12.052.
[38] G. Wu, H. Zheng, X. Ma, C. Kutlu, and Y. Su, "Experimental investigation of a multi-stage humidification-dehumidification desalination system heated directly by a cylindrical Fresnel lens solar concentrator," *Energy Convers Manag*, vol. 143, pp. 241–251, Jul. 2017, doi: 10.1016/J.ENCONMAN.2017.04.011.
[39] H. Xu and Y. J. Dai, "Parameter analysis and optimization of a two-stage solar assisted heat pump desalination system based on humidification-dehumidification process," *Solar Energy*, vol. 187, pp. 185–198, Jul. 2019, doi: 10.1016/J.SOLENER.2019.05.043.
[40] S. F. Li, Z. H. Liu, Z. X. Shao, H. shen Xiao, and N. Xia, "Performance study on a passive solar seawater desalination system using multi-effect heat recovery," *Appl Energy*, vol. 213, pp. 343–352, Mar. 2018, doi: 10.1016/J.APENERGY.2018.01.064.

[41] G. Filippini, M. A. Al-Obaidi, F. Manenti, and I. M. Mujtaba, "Design and economic evaluation of solar-powered hybrid multi effect and reverse osmosis system for seawater desalination," *Desalination*, vol. 465, pp. 114–125, Sep. 2019, doi: 10.1016/J.DESAL.2019.04.016.

[42] K. Mohammadi, M. Saghafifar, K. Ellingwood, and K. Powell, "Hybrid concentrated solar power (CSP)-desalination systems: A review," *Desalination*, vol. 468, p. 114083, Oct. 2019, doi: 10.1016/J.DESAL.2019.114083.

[43] M. Mehrpooya, B. Ghorbani, and S. S. Hosseini, "Thermodynamic and economic evaluation of a novel concentrated solar power system integrated with absorption refrigeration and desalination cycles," *Energy Convers Manag*, vol. 175, pp. 337–356, Nov. 2018, doi: 10.1016/J.ENCONMAN.2018.08.109.

[44] A. Alkaisi, R. Mossad, and A. Sharifian-Barforoush, "A Review of the Water Desalination Systems Integrated with Renewable Energy," *Energy Procedia*, vol. 110, pp. 268–274, Mar. 2017, doi: 10.1016/J.EGYPRO.2017.03.138.

[45] A. Rafiei, A. S. Alsagri, S. Mahadzir, R. Loni, G. Najafi, and A. Kasaeian, "Thermal analysis of a hybrid solar desalination system using various shapes of cavity receiver: Cubical, cylindrical, and hemispherical," *Energy Convers Manag*, vol. 198, p. 111861, Oct. 2019, doi: 10.1016/J.ENCONMAN.2019.111861.

[46] M. B. Shafii, S. Jahangiri Mamouri, M. M. Lotfi, and H. Jafari Mosleh, "A modified solar desalination system using evacuated tube collector," *Desalination*, vol. 396, pp. 30–38, Oct. 2016, doi: 10.1016/J.DESAL.2016.05.030.

[47] C. A. Arun and P. C. Sreekumar, "Modeling and performance evaluation of parabolic trough solar collector desalination system," *Mater Today Proc*, vol. 5, no. 1, pp. 780–788, Jan. 2018, doi: 10.1016/J.MATPR.2017.11.147.

[48] X. Luo, Y. Zhu, J. Liu, and Y. Liu, "Design and analysis of a combined desalination and standalone CCHP (combined cooling heating and power) system integrating solar energy based on a bi-level optimization model," *Sustain Cities Soc*, vol. 43, pp. 166–175, Nov. 2018, doi: 10.1016/J.SCS.2018.08.023.

[49] H. H. Al-Kayiem and O. C. Aja, "Historic and recent progress in solar chimney power plant enhancing technologies," *Renewable and Sustainable Energy Reviews*, vol. 58, pp. 1269–1292, May 2016, doi: 10.1016/J.RSER.2015.12.331.

[50] P. Guo, T. Li, B. Xu, X. Xu, and J. Li, "Questions and current understanding about solar chimney power plant: A review," *Energy Convers Manag*, vol. 182, pp. 21–33, Feb. 2019, doi: 10.1016/J.ENCONMAN.2018.12.063.

[51] H. Rezk, A. S. Alsaman, M. Al-Dhaifallah, A. A. Askalany, M. A. Abdelkareem, and A. M. Nassef, "Identifying optimal operating conditions of solar-driven silica gel based adsorption desalination cooling system via modern optimization," *Solar Energy*, vol. 181, pp. 475–489, Mar. 2019, doi: 10.1016/J.SOLENER.2019.02.024.

[52] A. Kasaeian *et al.*, "Solar humidification-dehumidification desalination systems: A critical review," *Energy Convers Manag*, vol. 201, p. 112129, Dec. 2019, doi: 10.1016/J.ENCONMAN.2019.112129.

[53] M. M. Morad, H. A. M. El-Maghawry, and K. I. Wasfy, "A developed solar-powered desalination system for enhancing fresh water productivity," *Solar Energy*, vol. 146, pp. 20–29, Apr. 2017, doi: 10.1016/J.SOLENER.2017.02.002.

AUTHOR INDEX

Institute of Physics
Dirac House, Temple Back
Bristol BS1 6BE UK

ISSN: 1757-8981
ISBN 978-1-7138-9255-7

CURRAN ASSOCIATES INC.
proceedings.com